S268 Physical Resources and Environment
Science: a second level course

BLOCK

ENERGY 1

FOSSIL FUELS

Prepared for the Course Team by Andrew Bell

S268 Physical Resources and Environment

Course Team

Dave Williams (Course Chair)
Andrew Bell
Geoff Brown
Steve Drury
Chris Hawkesworth
Ian Nuttall (Editor)
Janice Robertson (Editor)
Peter Sheldon
Sandy Smith
Peter Webb
Chris Wilson
John Wright
Annemarie Hedges (Course Manager)
Charlie Bendall (Course Coordinator)

Production

Jane Sheppard (Graphic Designer)
Steve Best (Graphic Artist)
David Jackson (Series Producer, BBC)
Nicholas Watson (BBC Producer)
John Greenwood (Liaison Librarian)
Eira Parker (Course Secretary)
Marilyn Leggett (Secretary)
Lynn Tilbury (Secretary)

Course assessor

Professor Peter W. Scott, Camborne School of Mines.

Dedication

Professor Geoff Brown was a member of the Course Team when he was killed on the Galeras Volcano, Colombia, in January 1993. The Course Team dedicates S268 to his memory.

Acknowledgements

The Course Team gratefully acknowledges the contributions of members of the S238 course team (S238 *The Earth's Physical Resources*, 1984).

The Course Team also wishes to thank Sheila Dellow for careful reading of early drafts of the course material.

The Open University, Walton Hall, Milton Keynes MK7 6AA.

First published 1995.

Copyright © 1995 The Open University.

All rights reserved. No part of this publication may be reproduced, stored in a retrieval system or transmitted in any form or by any means without written permission from the publisher or without a licence from the Copyright Licensing Agency Limited. Details of such licences (for reprographic reproduction) may be obtained from the Copyright Licensing Agency Ltd, 90 Tottenham Court Road, London W1P 9HE.

Edited, designed and typeset by The Open University.

Printed in the United Kingdom by Eyre & Spottiswoode Limited, Margate, Kent

ISBN 0 7492 5148 4

This text forms part of an Open University second level course. If you would like a copy of Studying with the Open University, please write to the Central Enquiry Service, PO Box 200, The Open University, Walton Hall, Milton Keynes, MK7 6YZ. If you have not already enrolled on the course and would like to buy this or other Open University material, please write to Open University Educational Enterprises Ltd, 12 Cofferidge Close, Stony Stratford, Milton Keynes, MK11 1BY, United Kingdom.

Edition 1.1

S268block4part1i1.1

CONTENTS

1 An introduction to energy — 6
- 1.1 Energy, power and efficiency — 7
  - 1.1.1 Some basic concepts — 7
  - 1.1.2 Energy requirements of individuals — 10
  - 1.1.3 Energy demands of a modern society — 11
  - 1.1.4 World power demand — 12
- 1.2 Energy around us — 13
  - 1.2.1 Solar radiation — 13
  - 1.2.2 Tides — 15
  - 1.2.3 The Earth's internal heat — 16
- 1.3 Concentrating, storing and transporting energy — 16
  - 1.3.1 Concentrating energy — 16
  - 1.3.2 Storing and transporting energy — 17
- 1.4 Renewable and non-renewable energy supplies — 17
- 1.5 Summary of Section 1 — 19

2 Fossil fuels and the carbon cycle — 20
- 2.1 Natural stores of carbon — 20
- 2.2 The terrestrial component of the carbon cycle — 21
  - 2.2.1 Photosynthesis, respiration and decay — 22
  - 2.2.2 The origins of coal — 23
  - 2.2.3 The geological conditions under which coal forms — 24
  - 2.2.4 The physics and chemistry of coal formation — 28
  - 2.2.5 Impurities in coal — 30
- 2.3 The marine carbon cycle — 31
  - 2.3.1 Preservation of marine organic matter — 33
  - 2.3.2 Kerogen — 35
  - 2.3.3 Generating petroleum from source rocks — 36
  - 2.3.4 Three examples — 39
  - 2.3.5 Impurities in petroleum — 39
- 2.4 Gas from coal — 40
- 2.5 Generating and concentrating carbon — 41
  - 2.5.1 Volcanoes and the carbon cycle — 42
  - 2.5.2 Productive environments, present and past — 42
  - 2.5.3 Preserving carbon-rich rock sequences — 43
  - 2.5.4 The fossil fuel bank — 43
- 2.6 Summary of Section 2 — 44

3 Finding and extracting coal — 47
- 3.1 Exploring for coal — 47
  - 3.1.1 Mapping coalfields — 47
  - 3.1.2 Drilling — 49
  - 3.1.3 Borehole logging — 49
- 3.2 Winning coal in former times — 51
  - 3.2.1 Early approaches — 51
  - 3.2.2 Pillar-and-stall working — 52
  - 3.2.3 The problems of water and air, and the solutions — 52
  - 3.2.4 The tools of the trade — 54
- 3.3 Modern coal production — 54
  - 3.3.1 Surface mining — 54
  - 3.3.2 Underground mining — 56
  - 3.3.3 Geological problems in coal mines — 58

3.4	A modern coalfield case study: the Selby complex in North Yorkshire	60
3.4.1	Some general considerations	60
3.4.2	The potential prospect at Selby	61
3.4.3	Exploration stage	62
3.4.4	Mine design and planning stage	63
3.4.5	Mining in the Selby complex	64
3.5	**Environmental aspects of coal mining**	65
3.5.1	Opencast mining	65
3.5.2	Deep mining	66
3.6	**Coal reserves worldwide**	68
3.6.1	UK coal reserves	71
3.6.2	Europe's coal reserves	72
3.6.3	World coal reserves	74
3.6.4	The life expectancy of world coal	75
3.7	Substitution and economics of coal production in Britain in the early 1990s	76
3.8	Summary of Section 3	79

4 Finding and extracting petroleum — 81

4.1	Fluids and petroleum migration	81
4.2	Reservoirs, seals and traps — the play concept	82
4.2.1	Reservoir rocks	83
4.2.2	Seals	84
4.2.3	Traps	84
4.2.4	The petroleum play	88
4.3	**Exploring for oil and gas**	90
4.3.1	Regional geophysical exploration	90
4.3.2	Seismic surveys	92
4.3.3	Exploration drilling	94
4.4	**Petroleum production**	95
4.4.1	Appraisal	95
4.4.2	Production techniques	96
4.5	A modern oilfield case study — the Brent field in the northern North Sea	98
4.5.1	Geology	98
4.5.2	Geophysics	100
4.5.3	Production strategy	100
4.6	**Environmental aspects of petroleum exploitation**	101
4.6.1	The environmental impact of transporting oil and gas	101
4.7	**Unconventional sources of petroleum**	102
4.7.1	Solid petroleum	102
4.7.2	Orimulsion	103
4.7.3	Gas from the deep-sea bed	103
4.8	**Oil and gas reserves worldwide**	104
4.8.1	Estimating reserves	105
4.8.2	UK petroleum reserves	106
4.8.3	World oil	108
4.8.4	World gas	110
4.8.5	Reserve predictions from discovery and production curves	110
4.9	Summary of Section 4	113

5 The future for fossil fuels — 116

5.1	World energy use and fossil fuel lifetimes	116
5.1.1	A case study — the lifetime of oil	118

5.2	Video review: Energy at the Crossroads	120
5.3	Burning fossil fuels	122
5.3.1	Burning pure hydrocarbons	122
5.3.2	Burning the chemical impurities in fossil fuels	122
5.4	Fossil fuels and atmospheric pollution	124
5.4.1	Sulphur dioxide and acid rain	124
5.4.2	Ozone	126
5.4.3	Carbon dioxide, methane and global warming	127
5.5	What about the alternatives?	132
5.6	Summary of Section 5	133

Objectives — **135**

Answers to Questions — **136**

Answers to Activities — **144**

Acknowledgements — **149**

1 AN INTRODUCTION TO ENERGY

Ten thousand years ago, when humans colonized Britain after the last Ice Age, they relied on hunting and gathering for food, and burning wood to keep warm. Their exact energy demands can at best only be estimated but to survive they probably needed some 20 MJ (megajoules) per person per day, about as much energy as it takes to run a couple of ordinary light bulbs continuously. By the time that Julius Caesar visited Britain 8000 years later, people travelled on foot, on horseback or in wheeled carriages, many ate well and some even had central heating in their homes. Wood was still their chief fuel, so despite their comparative luxury their average energy demands had probably not risen above 40 MJ each per day.

Thirteen hundred years after the Romans had left, Abraham Darby first used coal to smelt iron (Block 1, Section 1.3; Audio Band 1: *The Great Iron and Steel Rollercoaster*), but these newly industrialized humans still used only some 50 MJ each per day. Today, the average daily energy consumption per person worldwide is over three times that amount, about 175 MJ.

World population rose from some 5 million ten thousand years ago, through 200 million in Roman times, some 1 billion in Darby's time to roughly 5.5 billion today (Activity 2 in Block 1). This rapid increase in population, together with a sharp increase in the demand we each have for energy, has led to a dramatic rise in global energy consumption (Figure 1). **Fossil fuels** (coal, oil and gas; introduced in Block 1, Section 4) largely satisfy this staggering energy demand.

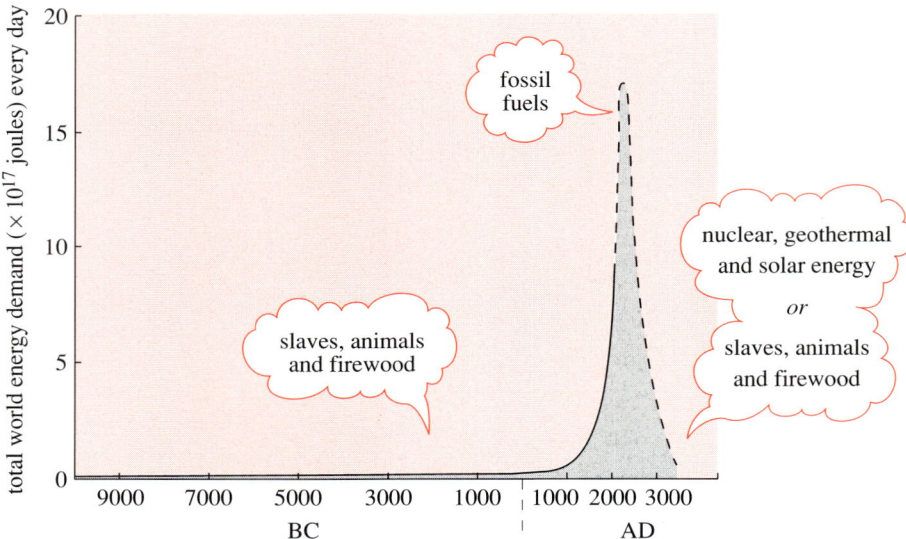

Figure 1 A graph of estimated global daily energy demand since the last Ice Age, some 10 000 years ago. The graph is extended to speculate what might befall humans after the era of fossil fuels ends.

In the natural world, energy supply limits population numbers. For example, large, fierce carnivores at the top of the food chain are very much outnumbered by their prey, because hunting demands a great expenditure of energy. If numbers of hunters increased dramatically, the amount of energy available to each individual (from their prey) would be substantially reduced and they would be unable to hunt successfully. But humans have broken free of such energy constraints. We provide for ourselves many times our basic energy needs through the cultivation of crops, the domestication of animals and by harnessing the energy sources described in this Block. This amounts to a fundamental change in the balance between life and earth processes, with consequences which we are only just beginning to understand. Our current position has resulted directly from exploiting fossil fuels over the past two hundred years — and our future will be closely linked to the future of energy generation, as Figure 1 so graphically shows.

1 An introduction to energy

Understanding energy generation means investigating both fossil fuels and renewable energy resources from a scientific and technological point of view. This Block establishes the uses, limitations and consequences of generating energy from each source available to our society.

Before we look at generating energy from fossil fuels, let us first take a closer look at energy itself. **Energy** is formally defined as *the capacity to do work*. Energy exists in many forms in the natural world: heat, light, sound, mechanical energy through movement (**kinetic energy**) or through position (gravitational or **potential energy**), electricity, chemical energy and nuclear energy. They all enable work of some form to be done. For example, sound causes your eardrum to vibrate, falling water can turn a water wheel, and so on.

Individuals and industry both need energy. Individuals must eat food to provide the body with heat energy for warmth and mechanical energy to walk, run and work, but our lives would be very unpleasant without external sources of energy. Even the most basic of human societies burn wood for cooking and for keeping warm, or use animals for transport and cultivation. Modern industrial societies use energy constantly and copiously: for manufacturing, for agriculture, for transport and for a host of other uses. Simply looking about us emphasizes how dependent we all are on external energy sources.

Question 1

Wherever you are at present, look around and decide how many of the materials around you required either (a) no energy at all in their formation, or (b) no human-made energy in their formation.

All the Earth's physical resources depend on energy to some extent for their extraction, processing and transport. If energy were to be cheap and plentiful, then grades of ore that would otherwise be uneconomic could be worked, stone could be quarried and transported anywhere, or water could be purified. In effect, the ability to extract and use the Earth's physical resources depends on whether there is a ready supply of energy at the right price. If there were an inexhaustible supply of cheap energy we could turn the entire stock of all resources into reserves.

1.1 Energy, power and efficiency

Terms like energy, power and efficiency are common in everyday speech. Politicians have power; children have energy; a good worker may be efficient. There are, however, rigorous scientific definitions which give a specific meaning to each of these terms, and it is important that you understand them.

1.1.1 Some basic concepts

Energy is required to do work, such as moving an object or making the molecules of a substance vibrate faster. All forms of energy are measured in **joules** (J), where 1 joule is the work done when a **force** of 1 newton moves its point of application 1 metre. A joule is the same amount of energy irrespective of which form (light, sound, electrical, and so on) that energy takes.

Actually, a joule is a rather small amount of energy. To lift a 1 kg bag of sugar about a metre from the floor onto a table takes almost 10 joules.

potential energy (J) = mass (kg) × height (m) × acceleration due to gravity (m s^{-2})

$= 1 \times 1 \times 9.81$ J

$= 9.81$ J

A weightlifter lifting 200 kg to a height of 2 m expends 4000 J of energy. Because a joule is such a small unit, it is more usual to express energy in *kilojoules* (1 kJ = 10^3 J), *megajoules* (1 MJ = 10^6 J) or *gigajoules* (1 GJ = 10^9 J). 1 kJ of energy can light a 100-watt electric light bulb for 10 seconds, or lift 100 bags of sugar onto a desk, or to bring 2.5 g of cold water to the boil, and 1 kJ of energy is released by burning (oxidizing) 0.03 g of pure carbon. The amount of energy is the same; the effects are strikingly different.

In developed countries, we need and use energy in many different forms. Most homes are supplied with electricity — a convenient domestic energy, readily convertible into heat (internal energy), light or movement. The majority of our homes are also supplied with gas — stored chemical energy which releases internal energy (heat) when it burns.

● Which energy changes take place when we use an electric kettle to boil water?

○ Electrical energy is first converted into movement of atoms to and fro in the metal element, which produces internal energy, or heat. Heat is conducted through the element into the water, making the water molecules vibrate, and heat up. Finally, heat energy is used converting water to steam.

The point to be grasped is that *changes* of energy from one form to another are commonplace in everyday life. All of the examples above involve energy being changed from one form into another.

Power is the *rate* at which energy is converted from one form into another. Being a rate, power is measured in joules converted per second (J s^{-1}), or more usually called **watts** (W). One watt is equivalent to one joule per second. Just as a joule is a small unit of energy, so a watt is a small unit of power. Larger quantities of power are measured in *kilowatts* (1 kW = 1000 watts or 10^3 J s^{-1}) or *megawatts* (1 MW = 1 000 000 watts or 10^6 J s^{-1}), or larger units like *gigawatts* (10^9 W) and *terawatts* (10^{12} W).

The point to note is that there is a difference between power and energy: energy is an amount and power is a rate of use.

Activity 1 How much electrical energy do we use at home?

Let's see how much electrical energy a typical household (yours) uses over various periods of time:

(a) You can use your electricity bills to calculate your household's average consumption of electricity over a period of weeks or months. Your bill should show both the billing period, often thirteen weeks, and the number of units of electricity used over that period. Each 'unit' on your bill is one *kilowatt hour* (kWh), which is actually an amount of energy, not of power as it first seems. Select a bill with an actual rather than an estimated reading if you can, and answer these questions. (You may need to look at two consecutive bills to get the exact billing period.)

i) How many joules are there in one kWh?

(ii) How many joules has your household used over this billing period? (Units × number of joules in one unit.)

(iii) What does that represent in joules each day?

(iv) What is your average power consumption over this period?

(b) Reading your electricity meter every few days will show your actual daily rate of electricity consumption. Read your meter now and fill in today's date, the time you took the reading, and the exact meter reading in the first row of Table 1 below. Tomorrow, and again after a few more days (when you've finished studying this Section for example), read the meter again and complete the relevant rows in Table 1. What is your energy consumption each day and your daily power use over the total time period?

Table 1

	Date	Time	Electricity meter reading/kWh
Start			
After 24 hours			
Finish			
Totals		(days) (hours)	

(c) When you have completed Table 1, write down some reasons why your average electricity consumption in joules per day should be different over a period of weeks (part a of this Activity) compared with the consumption over a period of days (part b of this Activity).

Energy can neither be created nor destroyed. Every application that 'uses' energy is in fact converting one form of energy into other forms. Some of this energy conversion is useful, some is not. Your kettle boils water, but it may also 'sing' (sound energy) and heat up itself. A **hydroelectric power** station turns the kinetic energy of moving water into electricity through generators, but at the same time friction between the moving parts of the generators produces heat and sound. The heat and sound are smaller, less useful, packages of energy.

The concept of **energy efficiency** quantifies the amount of useful energy retained during conversion. The efficiency of any energy-changing process is the ratio of useful energy output from that process to energy input into the process. Efficiency is a ratio, a dimensionless quantity usually expressed as *per cent*. A perfect energy-changing machine would get out as much useful energy as was put in, so it would be 100% efficient, but sadly such a machine does not yet exist. Real machines are much less efficient. For example, most of Britain's thermal power stations are only about 33% efficient because, like all thermal energy conversion machines, their efficiency is limited by the ratio of the outlet to inlet temperature of the turbine.

Exactly how useful is a joule of energy or a watt of energy change to us? A supply of 1 kilowatt (one thousand joules of energy changed each second) will run a one-bar electric fire, or ten light bulbs. Working really hard, the human body can just about deliver 1 kW for a very short period of time. This is the effort needed to lift a stone block weighing 100 kg from the ground onto the top of a metre-high wall, taking only a second to do it.

Activity 2 Power ratings

(a) My kettle has a 2 kW rating. How long will it take this kettle to bring a cup of water to the boil? (Assume the cup holds 200 g of water which has an initial temperature of 20 °C.)

(Note: It takes 4.18 J to raise the temperature of 1 g of water through 1 °C.)

(b) Most kettles are rated nominally, rather than accurately. Why not perform an experiment to work out the exact power rating of your own kettle?

Here are some tips and hints. You'll need to raise the temperature of a known quantity of water through a known temperature interval, and also to measure the time it takes to do this. Use large quantities to get reasonably accurate results. Find out the temperature of your tapwater and the temperature at which water boils, either by using a thermometer (range at least 0–100 °C) or, less accurately, by making reasoned guesses at these values.

Now work out how many joules have been used in bringing the water just up to the boil, and from the time it took you can work out how many watts the kettle actually supplies.

Either design your own experiment or, if you prefer, follow the 'recipe' set out in the answer to this Activity. Have a go, it's easier to do it than to read about it.

1.1.2 Energy requirements of individuals

In our society, the daily energy demands of each individual are far greater than the energy each one of us needs to run our own bodies. We are familiar with measuring the energy content of the food we eat (a stored form of chemical energy) in *kilocalories*. On average we each eat between 2000 and 4000 kilocalories of food a day, partly depending on the type of activities we do. 4000 kilocalories represents the food intake needed to sustain a hard day's physical work, but it is also equivalent to 16.7 MJ per day, about 190 joules per second or 190 W. Working hard, we convert in our bodies each day about the same amount of energy as it would take to run a large light bulb all day long.

How do our bodily energy needs compare with the amount of energy we use in our homes? Try Question 2, and combine the answer with your data from Activity 1 to get a feel for the answer to this question.

Question 2

The occupiers of one particular all-electric maisonette do not have to pay their own electricity bills. They use their electrical appliances as follows.

There are two 2 kW electric fires on all day (12 hours), but not at night.

A 650 W microwave is in use for 1.5 hours in total each day.

A 500 W fridge runs all day (24 hours).

The shower is used (5 kW) for total of 1 hour each day.

The washing machine is on (2.5 kW) for total of 1 hour each day.

Both rooms are lit, with 150 W bulbs in each room, for 24 hours every day.

What is (a) the daily energy requirement and (b) the average power requirement for this maisonette?

The total power use each day of a UK home can be many times the amount of energy we use to 'run' our bodies. The electricity demands alone of the home mentioned in Activity 1 are three times a human body's maximum energy requirements; the maisonette in Question 2, where the occupiers consume more energy than most of us, uses sixteen times this amount.

1.1.3 Energy demands of a modern society

Developed countries, with industrial as well as domestic demands, use energy in vast quantities and at alarming rates. In 1992, UK energy consumption (all forms, excluding food) was the equivalent to 9.14×10^{18} J of electricity. On average, that's about 1.6×10^{11} J for each of the 57 million people in the country. In the UK, about one-fifth of the country's energy requirements is used in the home; industry, commerce and transport account for the other four-fifths.

Is Britain a 'typical' country? Well, the energy demands of any country depends both on number of people who live there and on its degree of industrialization. Industrialized countries with smaller populations than Britain have similar energy requirements *per person* to Britain's. The Republic of Ireland (population 3.5 million, or 6% of Britain's) and New Zealand (population also 3.5 million) have total energy demands (1992) of 4% and 6% of that of Britain respectively.

Developing countries with larger populations than Britain but with a different type of industrial base, like India, use less energy per person than Britain. In 1992, India used energy equivalent to 92% of British requirements, but the population of India, at 766 million, is 13.5 times the British figure. This represents a yearly energy use of about 1.1×10^{10} J per person.

- So how does India's 1992 energy use per head compare with Britain's?

- India's energy use per head, 1.1×10^{10} J, divided by Britain's, 1.6×10^{11} J, gives a ratio of 0.0688. India's per capita energy use is 7% that of Britain.

Some industrialized countries with larger populations than Britain use much more energy per person than Britain. In 1992 the USA used a staggering 8.47×10^{19} J, or over one-third of the energy used in the world that year. With a population of 239 million, this amounted to 3.5×10^{11} J per person: about twice the British and 32 times the Indian demand.

The total world energy demand for 1992 was 3.37×10^{20} J. With a current world population of some 5.5 billion (Block 1), the energy demands of each world citizen in that year were, on average, 6.13×10^{10} J.

- Express the Indian, British and American per capita energy demands (excluding food) in terms of the global average.

- The calculation can be done by dividing each country's per capita energy demand in joules by the average in joules. India is 1.1×10^{10} divided by 6.13×10^{10}, or 0.18 of global average. Similarly, Britain is 2.6 times average and USA is 5.7 times average.

Question 3

Figure 2 shows *on a logarithmic scale* some of the energy requirements discussed so far in the Block.

(a) Add to this energy scale the daily energy requirements of the maisonette described in Question 2; the 1992 energy demands of New Zealand, UK and USA; and the total world energy demand in 1992.

(b) In hypothetical energy calculations (100% efficiency), one tonne of oil equates approximately to 4.32×10^{10} J of electricity. This energy figure is called *one tonne of* **oil equivalent**. Add the point representing the ideal 'one tonne oil equivalent' to the scale in Figure 2.

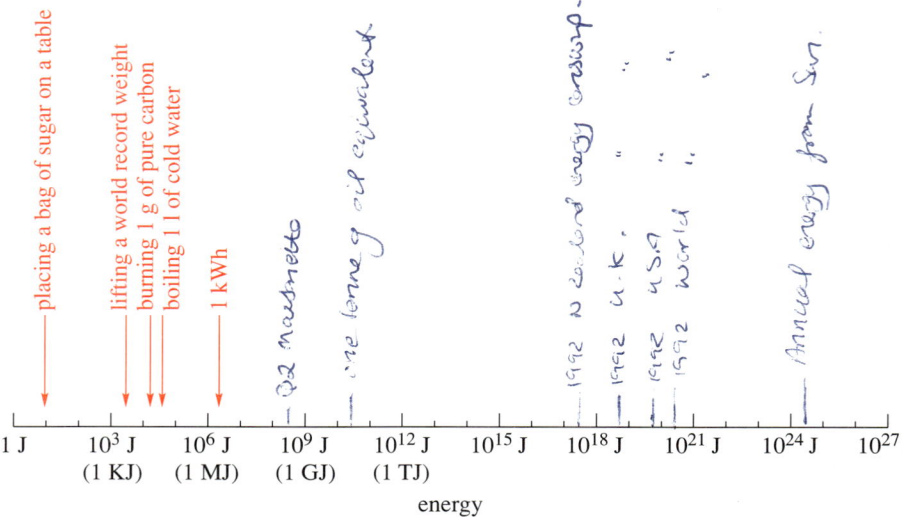

Figure 2 Some typical energy requirements. Note that the energy scale is logarithmic.

Do we generate energy efficiently?

According to the *BP Statistical Review of World Energy* (from which these 1992 figures came), 'one million tonnes of oil produces about 4×10^9 kWh of electricity in a modern power station.' How efficient is that?

We know from Question 3 that one tonne of oil ideally produces 4.32×10^{10} J of electricity. We are told that one tonne of oil burned *in this power station* produces 4×10^3 kWh of electricity, which is 1.44×10^{10} J (1 kWh is equivalent to 3.6×10^6 J). The efficiency of the power station is given by energy output divided by energy input, or $1.44 \times 10^{10} / 4.32 \times 10^{10}$. This 'modern power station' is therefore only 33% efficient.

1.1.4 World power demand

We can calculate the average power demand that matches the energy demand figures in Section 1.1.3. Since the world used 3.37×10^{20} J of energy in 1992, we could write that as 3.37×10^{20} J y^{-1}. A year contains 3.15×10^7 seconds so the average power use of the world in 1992:

$$= \frac{3.37 \times 10^{20}}{3.15 \times 10^7} \text{ J s}^{-1}$$

$$= 1.07 \times 10^{13} \text{ W}$$

This, the global power requirement, is an important figure. Many environmentalists quote the value of 10 TW (ten terawatts, or 10×10^{12} W) as an average global power demand for the early 1990s.

● If the world population is about 5.5 billion, how much power does each person in the world require, on average?

○ Dividing 10 TW (10^{13} W) by the current world population of about 5.5×10^9 gives an *average* rate of just under 2 kW per head worldwide.

The average figure of almost 2 kW per person worldwide is substantially less than the 5 kW each UK citizen demands. If every individual in the world were to demand as much energy as we in the UK use, the world's energy supply industries would be required to increase their output to 27.5 TW. [25 TW] Even more daunting is the prospect of exponential world population growth. If world population grows, as predicted, to over 10 billion in the year 2035, the figure would be 50 TW, or five times as much as today.

1.2 Energy around us

The natural environment itself is full of energy from other sources. Standing on a cliff top on a bright spring day you can feel the warmth of the Sun and the freshness of the breeze and hear the crashing of breaking waves below. Some natural phenomena are very energy intensive: a discharge of lightning releases 10^9 J and the 1906 San Francisco earthquake released an estimated 3×10^{17} J of energy.

To get the total global energy supply and demand into proper perspective we need to consider the contribution made to the Earth's energy supply by different *natural* energy sources. By far the most important supplier of energy on Earth is the Sun, but some energy comes from the gravitational attraction between Sun, Moon and Earth, and some from the Earth's own internal heat.

1.2.1 Solar radiation

Over 99.9% of the energy on Earth comes from the Sun. In common with other stars, the Sun is a vast nuclear powerhouse producing heat, light and other types of electromagnetic radiation from energy released by nuclear reactions. The Sun's power output is enormous, some 10^{26} W, but only a tiny percentage, about 17×10^{16} W, reaches the Earth. A third of this is reflected back into space (Figure 3).

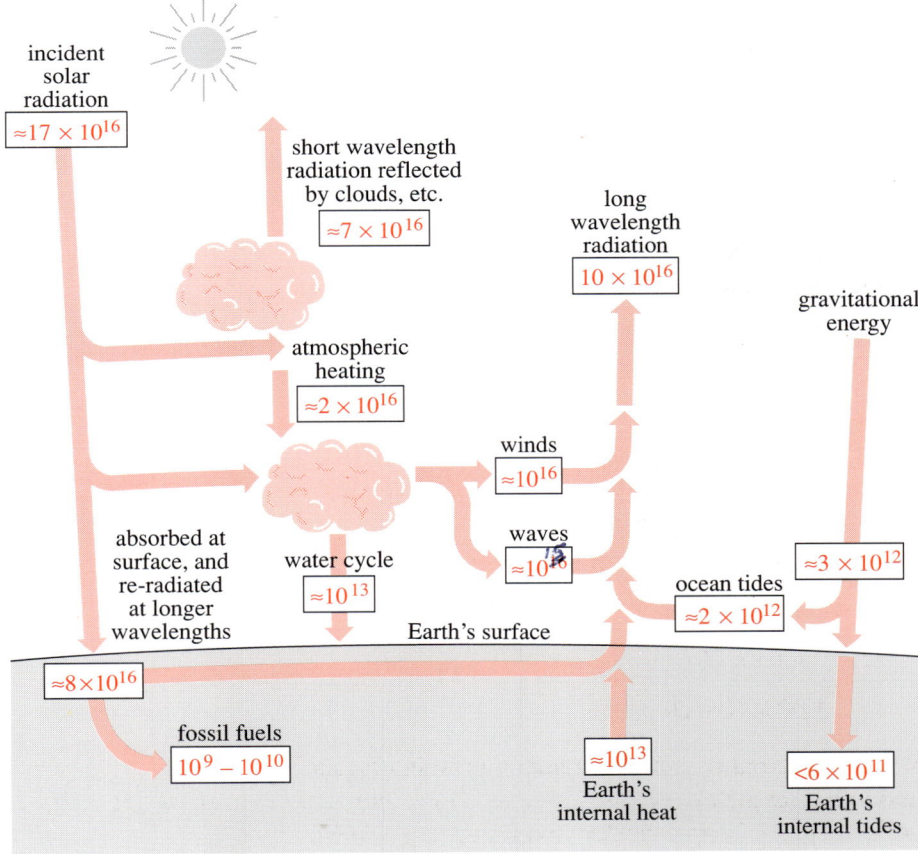

Figure 3 A flow chart detailing the amount of change of energy within the Earth's natural energy systems, measured in watts.

Block 4

Question 4

How much *energy* reaches the Earth (surface and atmosphere together) from the Sun each year? Add this number to Figure 2.

If all the solar energy that reaches the Earth could be harnessed, humans' current needs would be supplied thousands of times over. The reason it cannot is, as we shall see later, one of the most fundamental factors in any energy debate.

- ● Very approximately, what is the average solar power supply in watts per square metre over the Earth's illuminated surface? Assume that the Earth is a sphere with a 6400 km radius and that it is uniformly lit (which it is not).

- ○ If the Earth approximates to a sphere with a 6400 km radius ($r = 6.4 \times 10^6$ m), it has a surface area of $4\pi r^2$, of which only half (the 'daylight' side) receives sunlight at any one time. The illuminated area is therefore $2\pi r^2$, or $(2 \times 3.14 \times 4.1 \times 10^{13}) = 2.57 \times 10^{14}$ m^2. The total power supplied by the Sun that reaches the Earth's surface $(17 \times 10^{16} - 7 \times 10^{16}$ W) averaged over this area is about 390 W m^{-2}.

This very generalized calculation, which doesn't allow for latitude variations, confirms that solar illumination contains enough energy per square metre to provide intense lighting; something the Sun does very effectively every day in all but the most distant polar regions.

When it reaches the Earth's surface, solar radiation is potentially available as an energy resource either by humans using solar cells or panels, or naturally through plant and animal growth. Some 9×10^{12} W of absorbed solar radiation is transferred back into the atmosphere through the water cycle. Some of this energy reappears in usable form (from a human point of view) when water eventually returns to the surface as precipitation. This energy is largely dissipated as frictional heat and sound during the return flow of the water to the oceans, but it can also be harnessed as hydroelectric power.

However if the energy demand of an average British household, say 3 kW, came from solar power alone, even on a bright day every house in Britain would need about 80 m^2 of 10% efficient solar panels.

- ● Is solar power the answer to world energy supply? California has an area of about 405 000 km^2 (4.05×10^{11} m^2). Assuming that the Sun shines there for 12 hours each and every day of the year, what percentage of California's land surface area would need to be covered by standard solar panels (10% efficient) to supply the 1992 energy demands of the whole USA?

- ○ The USA used 8.47×10^{19} J of energy in 1992. A 10% efficient, metre-square solar panel would (from the question above) supply 39 J every second it was operative, and assuming that it operated for 1.57×10^7 seconds (half a year, but far more than ever could be achieved), one such panel would supply 6.1×10^8 J. The USA would therefore need 1.39×10^{11} such panels (8.47×10^{19} divided by 6.1×10^8). Since the area of California is 4.05×10^{11} m^2, almost 35% of the Californian land area would have to be covered by solar panels to meet this requirement. The assumptions we have made are so unrealistic that the figure would be much higher, were a 'solar power only' scheme to be implemented.

What happens to solar energy in the natural environment? Because the Earth is curved, the heating effect of the Sun is greater in equatorial latitudes than at the poles. Coupled with the Earth's rotation, this produces winds which blow between belts of high and low atmospheric pressure (Figure 4). About 10% of the Earth's absorbed solar power (10^{16} W) gives rise to winds. Much of the energy of winds is dissipated as heat, but some 10% of 'wind energy' is converted into water waves through frictional effects at the sea surface.

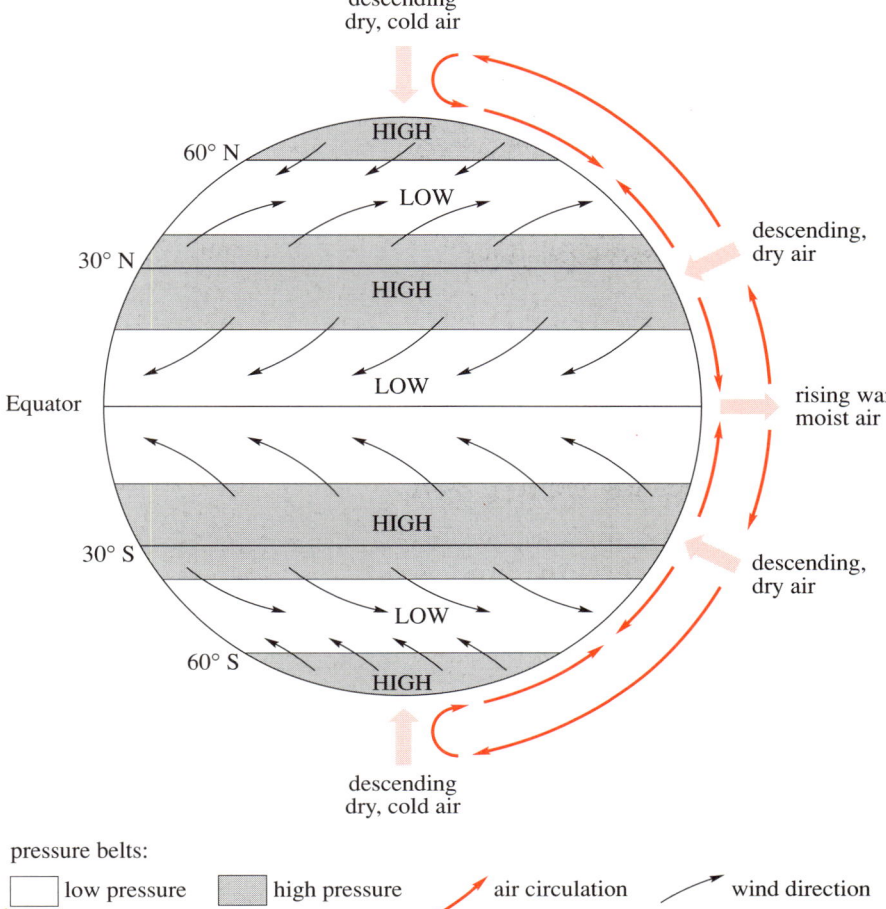

Figure 4 The main belts of high and low pressure together with the associated wind systems (arrows) caused by the combined effects of differential solar heating between the Equator and the Poles, and the Earth's rotation. This diagram assumes that the Sun is overhead at the Equator (as happens at the spring and autumn equinoxes).

○ Could the world be supplied from wind and wave power one day?

● Theoretically, yes, but practically, no. Up to 10^{16} W could become available from winds, i.e. one thousand times human power demands. However, tapping wind power over even one-thousandth of the world's surface (some 510 000 km² or over twice the size of Britain) would be a colossal undertaking and the process would have to be fully efficient. Similar arguments apply to wave power, except that here a maximum of 10^{15} W is theoretically available (100 times annual demand), requiring twice the surface area of the Mediterranean Sea to be tapped at 100% efficiency.

1.2.2 Tides

Tides are caused by the gravitational pull of the Moon and to a lesser extent the Sun. Although tides affect all fluid bodies on Earth in some measure, including some parts of the solid Earth itself, their main effect is on the seas and oceans. Ultimately the kinetic energy of tides is converted into heat, mainly through friction between water and the sea bed. Tides can be exploited as an energy resource, and the total amount of power available can be calculated from a knowledge of the Earth–Moon–Sun system. At about 2×10^{12} W, it is many orders of magnitude less than the power of solar radiation (Figure 3) and less than 20% of the current power demand for human activities.

1.2.3 The Earth's internal heat

The natural occurrence of both volcanoes and hot springs shows that the Earth's interior is hot, producing molten rock at temperatures up to 1250 °C, and also superheated steam. However, these phenomena are mainly confined to several narrow and elongate zones along the world's active plate boundaries. Many measurements have now been made of the amount of heat flowing from the Earth's interior. Outside the distinctive zones mentioned above, heat flow varies from 40–120 milliwatts per square metre ($mW\,m^{-2}$), largely generated by the decay of long-lived radioactive isotopes within the Earth. The total power output from the Earth's interior, estimated at some 10^{13} W, is about 5000 times smaller than the total incident solar power (Figure 3).

- Could the world's requirement for energy come entirely from geothermal sources eventually?

- No. At 10^{13} W, this source is roughly only the same as current power demand for human activities, and it could not be fully tapped at 100% efficiency.

1.3 Concentrating, storing and transporting energy

The Earth is awash with energy; the Sun provides seventeen thousand times as much energy as humans use. Why then do we need to use any other energy supply?

1.3.1 Concentrating energy

As far as human needs are concerned, there is a marked difference between 'dilute' and 'concentrated' energy. Water vapour in the atmosphere, for example, has considerable potential energy since its significant mass (15×10^{12} tonnes; Block 3 *Water Resources*, Section 2.1) is held high above the Earth's surface. But this potential represents a very dilute form of energy; falling rain could not turn a water wheel. It is only when energy can be concentrated that it can be put to good use.

Some forms of energy are relatively difficult to concentrate, others are easier. The energy contained in moving air is rather difficult to concentrate; windmills and wind farms have to be sited where natural factors enhance wind speed and constancy. The potential energy of rain is naturally concentrated and held in mountain lakes; we concentrate this energy artificially when rainwater is stored in a reservoir.

In the past we have chosen to generate our electrical energy from the most concentrated sources available to us. Coal, oil and gas are so valuable because, as we shall see, they represent concentrated forms of the solar energy that reached the Earth millions of years ago.

The ultimate form of concentrated energy is matter itself. When Einstein formulated his famous equation $E = mc^2$, he showed that mass and energy are proportional to one another (the constant linking the two is the square of the speed of light). Scientists now realize that the conversion of matter (with

1 An introduction to energy

mass, *m*) into energy (*E*, in joules) is one of the most exciting future energy-producing possibilities.

- If we could design a perfectly efficient device that was able to convert pure matter into energy, how much matter would need to be converted each year to supply the world's current annual energy demands? The speed of light, *c*, is 3.00×10^8 m s^{-1}.

- Such a device would have to provide 3.37×10^{20} J each year. The mass of matter to be converted, *m*, is given by E/c^2, or $(3.37 \times 10^{20})/(9.00 \times 10^{16})$ kg. This is equivalent to about 3.74 tonnes of material.

The notion of generating all the world's energy from less than four tonnes of matter is pure science fiction at present, but the point about concentrated forms of energy is made very well.

1.3.2 Storing and transporting energy

To be useful to us, energy must also be available where and when we want it, and in a form and in amounts we can handle. Severe weather systems concentrate natural energy wonderfully, but hurricanes associated with storm-force winds, driving heavy rain, thunder and lightning wreak havoc rather than top up our energy supplies.

Storing most forms of energy is very difficult. We have to re-heat our homes daily in the wintertime because they constantly lose heat, despite our attempts to insulate them. We cannot store light when the Sun goes down; we have to turn electricity into light until the Sun reappears. In fact, only two energy forms are truly storable. We can store almost indefinitely potential mechanical energy, or potential heat energy locked in certain chemicals. The first is the basis of clockwork and hydroelectric schemes; we know the second better as **fuel**.

Fuel is material with potential but as yet unreleased, concentrated energy. Wood was the major fuel before the Industrial Revolution, and remains the most important fuel for many non-industrial societies today. As industry develops, energy demands grow and fuels with higher convertible energy content per unit mass are needed. Modern energy supply is centred on the fossil fuels: coal, oil and gas.

A further advantage of fuels as energy sources is their transportability, so that conversion can take place on selected sites or in mobile units. Highly concentrated fuels require less energy to transport than dilute fuels, but since lots of energy can be released accidentally from badly handled concentrated energy sources, care has to be taken to ensure that transport is safe. For some applications, such as cars, generating energy from stored chemical energy has the advantage because of the ease of transport of small amounts of fuel.

Fossil fuels therefore represent extremely useful energy sources because they are concentrated, they store energy long term, and they can be simply and relatively safely transported.

1.4 Renewable and non-renewable energy supplies

Energy resources can, however, be considered in a completely different way — whether or not they are renewable. Some potential energy sources utilize energy released comparatively recently from the Sun. This energy is replenished continuously and naturally over a timescale of days to tens of years, regardless of whether or not it is turned by humans into other forms of energy. We refer to such energy resources as **renewable energy**.

Other potential energy resources are legacies of former solar power; coal is a good example. These are replenished naturally over a timescale of thousands or millions of years, so slowly that in human terms they may as well be considered as **non-renewable energy**.

● Which energy sources mentioned so far are renewable, and which are non-renewable on the scale of a human lifetime?

○ Energy sources such as solar, hydro, wind, waves, tides, and geothermal are continuously replenished, so they are renewable on this timescale. The fossil fuels are not being replaced on this timescale, so they are therefore non-renewable.

Actually, fossil fuels *are* slowly being renewed by the death, burial and decay of present plant and animal life, but at an extremely slow rate compared with our lifetimes. By the time the Earth's natural system has replaced all the fossil fuels we have already used, humans may not even exist as a species on Earth.

The distinction between renewables and non-renewables is one of timescale and energy concentration, but it is a distinction that is critical for human society. We can draw a parallel with our own domestic cash-flow. In life we spend money to live; the money might come from our pocket or our savings. Spending savings (a concentration of available money) will cause them to run low, but with careful money management they could perhaps be topped up from income (renewable funds). If a rich friend or relative leaves you a large inheritance, you might choose to keep this money (non-renewable) in your savings account, or to spend some of it in a way that benefits you.

Now think of renewable energy resources as *income*, and non-renewable energy resources as *inheritance*. We 'spend' the Earth's energy resources constantly for cooking, travelling, heating or cooling buildings, manufacturing and in many other ways. At present, modern industrial societies generate energy mostly from fossil fuels, using up an inheritance which has accrued from millions of years of investment of solar energy and internal heat. Much less energy is currently generated from renewables; i.e. from day-to-day energy 'income'. Unfortunately we have no practical way of replenishing our energy inheritance, of turning renewable energy into fossil fuels. The term 'non-renewable' is all too appropriate here.

Irrespective of any environmentally damaging effects of burning fossil fuels, such as atmospheric pollution or global warming, sooner or later our present energy generation policy will deplete our stock of fossil fuel. To stay solvent in energy *in the long term* we will ultimately have to change our day-to-day energy supply from non-renewable towards renewable energy sources, or return to a low-energy society (depicted in Figure 1). This Block aims to provide a scientific basis to understand some of the decisions we will need to make to achieve this.

Activity 1 reprise

Return now to Activity 1 and complete Table 1 by reading your meter again. You can then work out the amount of electrical energy your household has used over the time you've been reading this Section.

1.5 Summary of Section 1

1. Energy is the basis of modern society. Other physical resources can only be effectively extracted, processed and transported if there is ready supply of energy at the right price.

2. Energy (measured in joules) is defined as the capacity to do work, whereas power (measured in watts) is the rate of doing work or the rate at which energy changes from one form to another.

3. All conversions of energy are inefficient to varying degrees.

4. The Sun is by far the most important source of natural energy on the Earth. The solar radiation that reaches the Earth contributes to winds, waves, atmospheric water circulation, atmospheric heating and surface water evaporation, and to organic activity.

5. The gravitational pull of the Sun and the Moon combines with the Earth's axial rotation to produce tides, and the Earth is also internally hot. These are small but potentially exploitable sources of energy.

6. The world is plentifully supplied with solar-derived energy, but most of it is not in a sufficiently concentrated form to make it of use to a modern industrial society.

7. Fuels are of immense value because they are concentrated forms of energy that can easily be stored, transported and used at will.

8. Energy sources can be subdivided into renewables, like solar, wind and wave power, and non-renewables, like peat, coal, oil and gas. Renewables are effectively everlasting, but there is a finite amount of non-renewables.

Question 5

Of the following eight statements (a) to (h), six are correct and two are incorrect. Identify the two incorrect statements.

(a) When work is done, energy is not destroyed but its usefulness is diminished. Consequently, useful concentrated energy forms like electricity are degraded through work into less concentrated energy forms like heat and sound.

(b) Tidal energy depends on the gravitational attraction between the Sun, the Moon and ocean water, and the rotation of the Earth.

(c) Energy is the ability to do work whereas power is the rate at which energy is converted from one form into another.

(d) Although winds and waves are temporary features that individually never last for long, the power they produce is still regarded in resource terms as renewable.

(e) Large developed countries like the USA use many times as much energy per capita as large less-developed countries like India, and substantially more energy per capita than smaller developed countries like New' Zealand.

(f) The average global power demand in the 1990s is expected to be around 10 terawatts, but this figure is expected to fall into the following century as energy efficiency becomes commonplace.

(g) The heat flow up through the surface from the Earth's interior is approximately equivalent in total energy terms to the solar energy that penetrates the atmosphere and reaches the surface.

(h) Fossil fuels have formed the basis of western industrialization because they are extremely transportable, concentrated energy sources.

2 FOSSIL FUELS AND THE CARBON CYCLE

As we saw in Block 1, Section 1.2, carbon is the most sought-after element in the world economy. Besides fossil fuels, the only other major physical resource that contains significant carbon, in the form of calcium and magnesium carbonates, is limestone. **Graphite** and diamond are also made from carbon, but both occur naturally only in extremely small amounts compared with carbonate rocks and fossil fuels.

To understand why fossil fuels exist we need first to know where the major stores of carbon are on the planet and how, through organic activity, this carbon becomes fixed into both carbonate rocks and fossil fuels. We must then address what role Earth processes have played in concentrating carbon into fossil fuels.

2.1 Natural stores of carbon

The major natural stores of carbon can be identified and assessed by surveying the worldwide distribution of various different carbon-rich environments, and then sampling representative areas of each. Scientists disagree about some of the details of such assessments and therefore on the exact amounts of carbon in each store. However, several recent estimates fall near the values shown in Figure 5. Routine monitoring has ensured that the amount of carbon in the atmosphere is better understood than that in other stores.

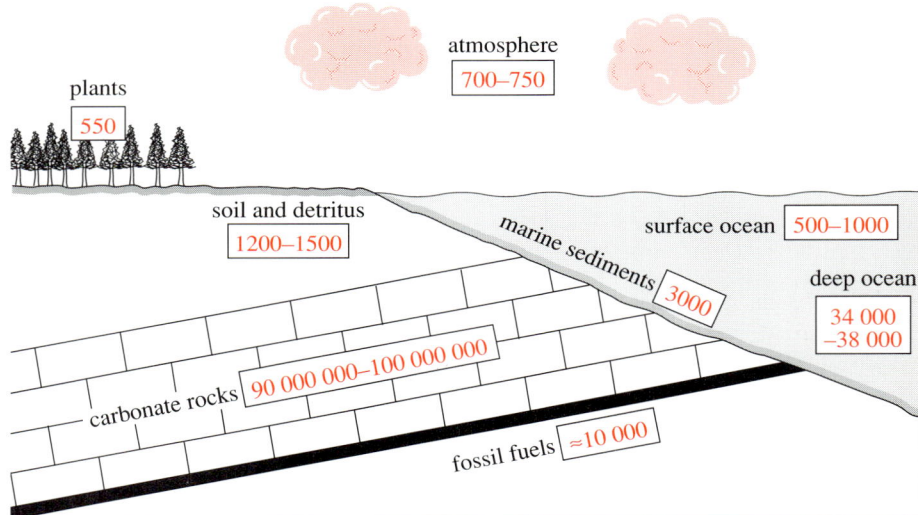

Figure 5 The eight major global stores of carbon. Amounts of estimated carbon are in thousand million (10^9) tonnes.

- ⚫ What percentage of carbon is held in carbonate rocks?

- 🟠 Carbonate rocks account for over 99.9% of natural carbon worldwide.

Carbon is continually being *exchanged* between the eight principal stores. As with total carbon amounts, exchange rates can be measured within each environment and the areas of each surveyed and quantified. Disagreements and uncertainties also exist.

Over periods of time measured in years to decades, there are significant exchanges of carbon between the stores shown in Figure 5. However, since most carbon is held as carbonates, the principal carbon exchange over the geological timescale (millions of years) is from the surface environments into limestones.

For our purposes two exchange systems can be distinguished (somewhat artificially because in practice the two are intimately linked):

1. the land-based or *terrestrial* exchange system by which carbon is exchanged between land plants and both the soil and the atmosphere;
2. the *marine* system which exchanges carbon within the oceans and between the oceans and the atmosphere.

Together they form the natural **carbon cycle**.

2.2 The terrestrial component of the carbon cycle

Figure 6 shows in billions of tonnes of carbon per year the *rates* of natural carbon exchange between the terrestrial system and the atmosphere.

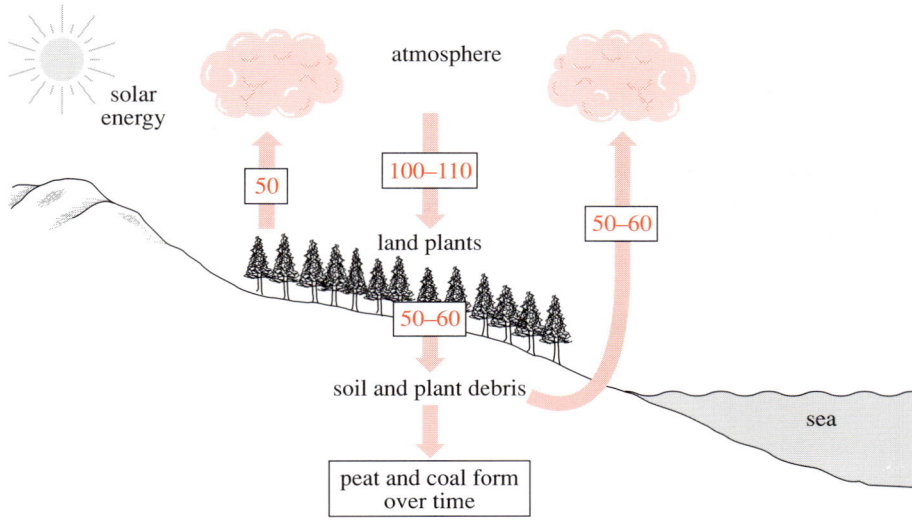

Figure 6 The terrestrial part of the carbon cycle, showing the rate of exchange of carbon between the land environments and the other carbon stores. Rates are in tonnes $\times 10^9$ of carbon per year.

○ Using the data in Figures 5 and 6, calculate whether there is a net movement of carbon into or out of the atmosphere, as far as the terrestrial carbon cycle is concerned. What is the average residence time of land-derived carbon in the atmosphere?

○ Figure 6 shows that $100-110 \times 10^9$ tonnes of carbon per year flows from land-based organisms, animals and the soil into the atmosphere, and $100-110 \times 10^9$ tonnes of carbon per year moves from the atmosphere into land-based organisms. As far as the terrestrial carbon cycle is concerned, there is *on average* a net balance of carbon flow. (We shall see how special conditions allow accumulations of plant material.) On average, about 15% of the 700×10^9 tonnes of atmospheric carbon (Figure 5) is removed and replaced each year. The residence time of terrestrial carbon in the atmosphere is therefore less than seven years, on average.

Each year land plants fix over one hundred billion tonnes of carbon; the most significant annual interchange of carbon in any part of the carbon cycle. How exactly is carbon exchanged between plants and the atmosphere?

2.2.1 Photosynthesis, respiration and decay

Green plants absorb solar radiation and use its energy to fuel **photosynthesis** — a chemical reaction in which carbon dioxide (CO_2) from the atmosphere is combined with water (H_2O) to form **carbohydrates** with the general formula $C_nH_{2n}O_n$. One of the simplest carbohydrates, glucose, has the chemical formula $C_6H_{12}O_6$, so in its simplest form photosynthesis can be represented by the balanced chemical equation:

$$6CO_2 + 6H_2O + \text{solar energy} = C_6H_{12}O_6 + 6O_2$$
$$\text{glucose}$$

The oxygen freed by this reaction is released by plants into the atmosphere. Carbohydrates act as a store of energy for plants and also for other organisms that eat plants. Organisms use oxygen from the air to react with the carbohydrates (and other substances) to liberate energy by a process called **respiration**. During respiration, waste carbon dioxide and water is returned back into the atmosphere. Expressed in the simplest chemical terms, the balanced reaction is:

$$C_6H_{12}O_6 + 6O_2 = 6CO_2 + 6H_2O + \text{energy}$$

which is the exact reverse of the photosynthetic reaction above.

Carbon exchanges or *fluxes* link the chemistry of the atmosphere with plant and animal chemistry. The carbon fixed by plants enables them to grow, but in addition much of it enters the food chain as either living or dead material. Living plants are eaten by herbivores which themselves may become food for carnivores. The dead material also provides food for the decomposers (bacteria and fungi) that live in plant detritus, in the soil, and on the rotting carcasses of dead animals. Almost all organisms return some carbon to the atmosphere through respiration, but by far the greatest contribution comes from the activities of the decomposers. The timescale by which this takes place is measured in months and years, so normally plant and animal material is not available to be preserved as fossil fuels.

However, if organic matter decays in an environment where oxygen supply is limited, carbohydrates cannot be broken down completely to form water and carbon dioxide. In this special oxygen-poor environment, a carbohydrate comparatively enriched in carbon may be produced. For example, within the waterlogged environment of a swamp, cellulose (a common constituent of plants) can be broken down according to the following reaction:

$$2C_6H_{10}O_5 = C_8H_{10}O_5 + 2CH_4 + 2CO_2 + H_2O$$
$$\text{cellulose} \quad \text{residue} \quad \text{methane}$$

The residue produced, $C_8H_{10}O_5$, is relatively enriched in carbon compared with the original cellulose ($C_8H_{10}O_5$ compared with $C_6H_{10}O_5$). This breakdown reaction releases **methane** (CH_4), as well as carbon dioxide and water. Methane is an organic compound containing carbon and hydrogen but no oxygen; one of a family of organic compounds known as **hydrocarbons**.

Under conditions that are oxygen poor or **anoxic** (lacking oxygen), not all the fixed carbon returns to the atmosphere as CO_2: some may be retained as carbon-enriched residue and yet more converted into hydrocarbons. Hydrocarbons represent a chemical half-way-house within the carbon cycle, which together with carbon-rich residues make carbon available to form the basis for fossil fuels.

Today, significant layers of decaying plant debris are found in extensive forests and jungles, but often these are oxygen-rich environments thanks to the constant reworking of decaying material by plants, animals and insects.

How peat forms

Peat is the compressed, soggy plant remains that characterize the bogs and swamps of poorly drained regions. There are extensive areas of peat in the upland districts of the British Isles, particularly in the Pennines, in Scotland and in Ireland. In these areas, peat has been extracted for use as a fuel. Peat bogs are not restricted to temperate latitudes or upland areas: they form in a wide range of settings from tropical rainforest to the high-latitude tundra regions.

The first requirement for the formation of peat is that large amounts of plant debris are available for preservation. Climatic conditions must exist that favour plant growth such as forests or jungle, although this is not restricted to the tropical regions of the Earth, as the forests of Siberia and Canada testify. In high latitudes, mosses rather than trees are the primary source of plant material. The climate must, however, be humid rather than dry.

For plant matter subsequently to be preserved as peat, the oxygen supply must be restricted. Swamps and bogs, where the water table is at or above the surface, provide such an environment since the soil pore-spaces are filled with water rather than air. Oxygen content is low because water contains much less dissolved oxygen than is present in air, and because the decay of plants matter consumes what oxygen is present faster than it can be replenished.

Extensive peat swamps can develop from plant colonization of shallow lagoons and lakes on low-lying coastal plains. Modern-day examples are found in the coastal plains of south-east USA, and in the densely forested swamps of South-East Asia. Boreholes through the Ganges delta in South-East Asia show a succession of peat beds with interbedded sand and clay, suggesting repeated alternations of plant growth and flooding, with the average land surface never far from sea level.

However, one modern environment that you are probably familiar with does contain plant material decaying in an oxygen-poor environment — the peat bog (see the Box 'How peat forms').

2.2.2 The origins of coal

Coal, like peat, forms from the remains of land plants. Coal contains over 50% by weight and more than 70% by volume of carbon. A polished block of coal has a prominent banded structure; the most common bands are made from continuous layers of soft, bright coal which breaks into angular pieces. Other shiny bands are made from thin shreds or a thin film of bright material in a fine-grained matrix. These two types of bright bands are separated by layers of dull, grey–black relatively hard coal (Figure 7). The dirtiness of coal comes from thin weak layers of charcoal-like carbon. Coal easily splits along these weak layers, which crumble to give coal its characteristic dusty black coating.

Figure 7 A typical sample of coal from a British coal mine. This comes from the Selby coalfield.

Microscopic inspection of samples such as that in Figure 7 shows that the bright coal represents single fragments of bark or wood, often showing a well-preserved cellular structure. The dull, grey–black relatively hard coal is an assemblage of crushed spores, shreds and grains of wood, and blebs of resin. The charcoal-like layers represent charred fragments of bark and wood.

Frequently, fragments of plants are preserved in sediments associated with beds of coal. These fragments include leaf fronds (Figure 8), branches and roots. Occasionally, fossilized tree stumps are found in sediments below coal seams, preserved in the positions they actually grew in.

Figure 8 Fossilized leaf (of the genus *Pecopteris*) in a Carboniferous Coal Measures deposit (roughly actual size).

The origin in land plants is also shown by the age of coal deposits. Coals are commonly found in rocks from Carboniferous times onwards, but Devonian coals are rare, and pre-Silurian true coals are never found. This coincides exactly with what we know about the evolution of land plants, which first appeared in Silurian times about 400 Ma (million years) ago, colonized the land surface rapidly through the Devonian and became abundant by Carboniferous times.

2.2.3 The geological conditions under which coal forms

Figure 9 is a diagrammatic summary of a typical vertical sequence of sedimentary rocks found in coalfields in many parts of the world. The following features are important:

1. Immediately beneath the coal seam is a **seatearth**, so called because it represents a fossil soil (Block 2 *Building Materials*, Section 4.6). Black carbonaceous markings, the rootlets of the plants, are common. Seatearths can be clays or sands, and are often a source of useful raw materials for the construction industry (Block 2, Section 4.2). Some are refractory clays (**fireclays**) which have high melting points; others are pure sandstones.

2. The coal seam itself is largely plant material with small but variable amounts of mud. The seam itself can vary in thickness from a few millimetres to tens of metres.

3. Immediately above the coal seam there may occasionally be a shale containing rare but distinctive marine fossils. This is known as a **marine band**, and it is unusual because most of the other fossil remains associated with coal-bearing sequences are normally of freshwater or land-based species, not species associated with the sea.

4   The muddy sediments which overlie the coal seam pass upwards into siltstones and then into sandstones. This sequence of rocks usually gets steadily coarser upwards, but variations are common. Fossils within this part of the sequence almost invariably indicate non-marine conditions. The sandstones and siltstones may show sedimentary features that indicate action of waves and currents in relatively shallow water. Some sandstones show evidence of river channels.

5   The sandstones usually pass upwards into seatearths and another coal.

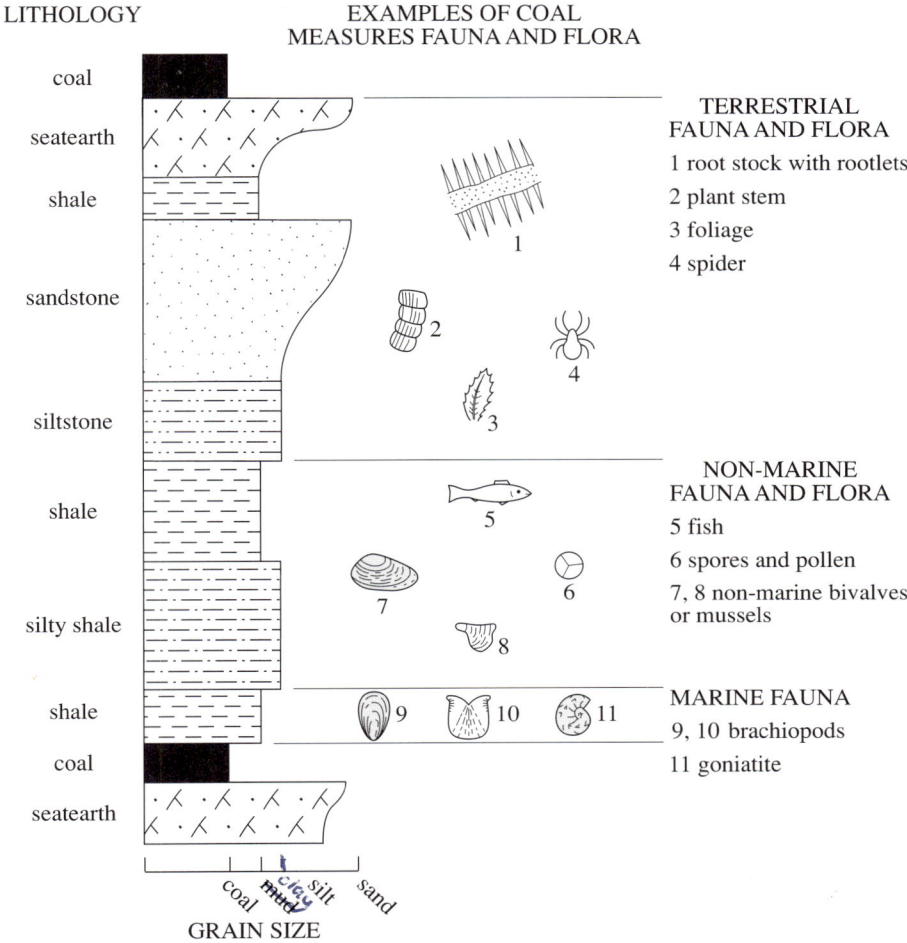

Figure 9 A vertical section through a typical sequence in a Coal Measures deposit, showing the sequence and variety of sediments, and some typical fauna and flora. As is conventional in diagrams like this, the vertical axis represents sediment thickness (here schematic) and the horizontal scale represents maximum size of grains within the sediment.

Taken as a whole, the sequence of rocks from coal through marine and non-marine shales to sandstones, seatearths and another coal seam tells us about the cycle of events that led to coal formation. Initially, flat low-lying sandy (or muddy) areas were covered by a forest of land plants. Rising water levels flooded the flat plains. Usually, vast freshwater lakes developed but occasionally the relative change of levels was sufficient to allow the sea to flood in. Rivers gradually encroached into the shallow lakes, building up deltas and sandbanks along the lake shoreline until eventually the sandbanks re-emerged to become recolonized by plants. The sequence is shown in Figure 10.

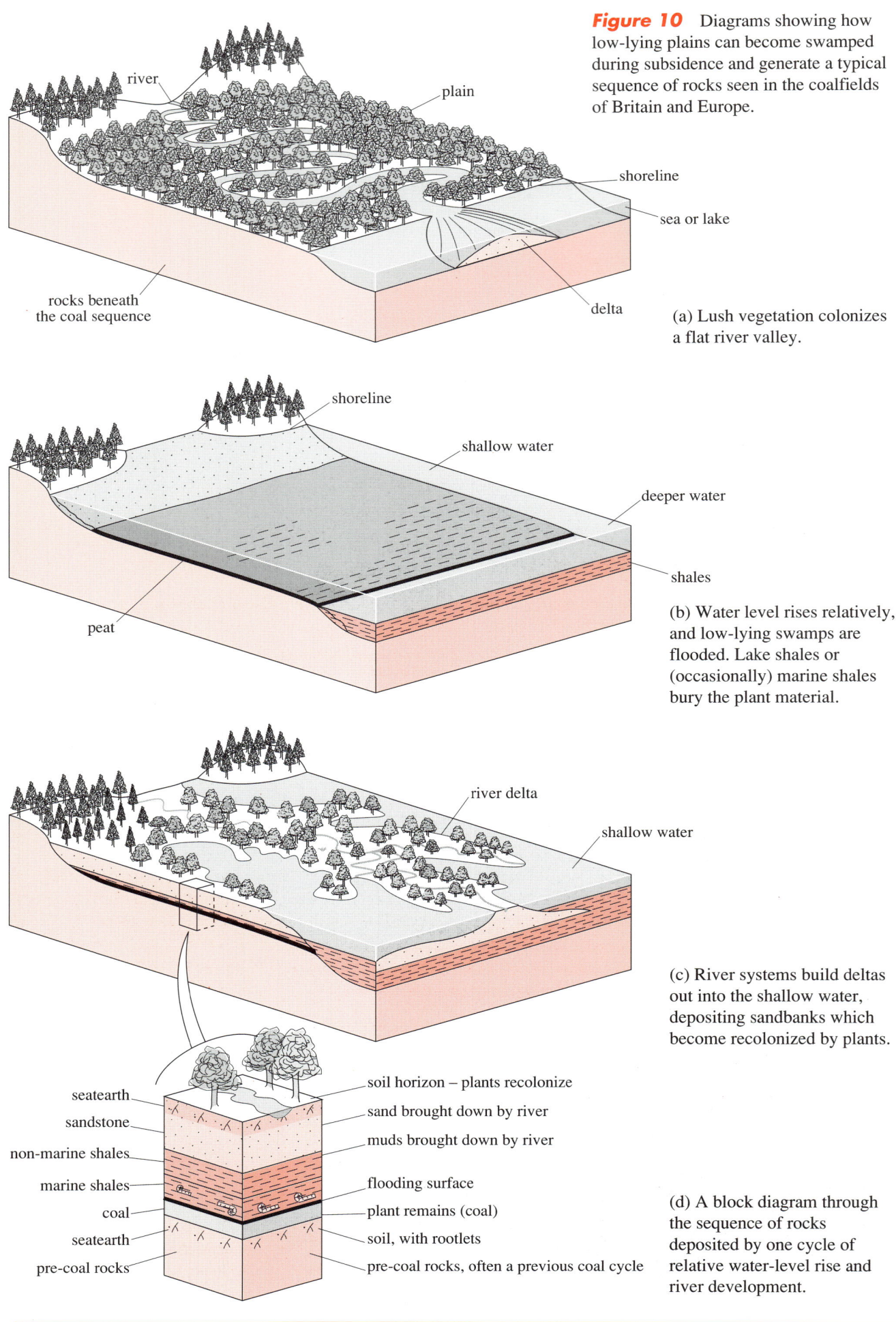

Figure 10 Diagrams showing how low-lying plains can become swamped during subsidence and generate a typical sequence of rocks seen in the coalfields of Britain and Europe.

(a) Lush vegetation colonizes a flat river valley.

(b) Water level rises relatively, and low-lying swamps are flooded. Lake shales or (occasionally) marine shales bury the plant material.

(c) River systems build deltas out into the shallow water, depositing sandbanks which become recolonized by plants.

(d) A block diagram through the sequence of rocks deposited by one cycle of relative water-level rise and river development.

These cyclic sequences of flooding by lakes or the sea, followed by re-deposition by river systems, were repeated time and time again throughout the late Carboniferous of Britain and Europe. While swamp conditions with little sediment input may have existed at one locality, sands, silts or muds were being deposited elsewhere. In many coalfields, it is found that seams either split into two or more beds separated by bands of sandstone or carbonaceous shale, or converge to join with an adjacent seam (Figure 11). The sediment sequence in coalfields can reach hundreds or even thousands of metres in thickness, even though all the sediments were deposited in shallow water. This is clear evidence that the coal cycles formed on parts of the Earth's crust that were subsiding.

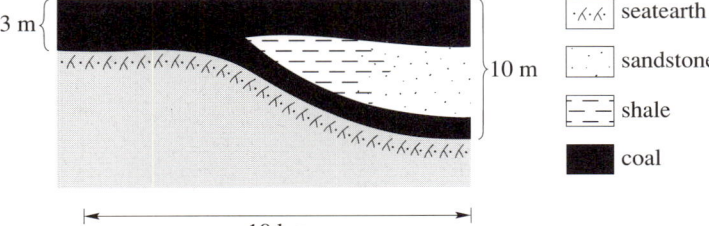

Figure 11 An example of splitting in a coal seam. Splits are caused by lateral variations in sediment deposition and plant colonization.

Regional geological studies show that individual cycles can vary considerably, both laterally within any one cycle and from one cycle to the next. However, some marine bands can be traced over vast distances, allowing sequences of cycles in different areas to be matched together.

Carboniferous coal swamps

During the late Carboniferous, coastal swamps developed over vast areas of Britain. Much of today's land area was an extensive, low-lying plain bordering the sea to the south. Any mountains that existed lay far away to the north in Scotland. Large river systems meandered southwards across these plains. The overall setting must have been similar to the present-day Irrawaddy delta in Burma.

At that time Britain lay in tropical latitudes, almost on the Equator. The low plains were covered by extensive forests: the Carboniferous equivalent of the present tropical rainforests. However, most of the trees were hollow, not solid, and more closely related to modern horsetails than to modern mangroves or mahogany trees. No flowers or birds existed, but insect and reptile life was abundant in the forests.

Tropical storms were also probably quite common — we have no reason to believe that storms and tropical cyclones were any less frequent in Carboniferous times that they are today. Such storms would devastate vast areas of forest, reducing acres of trees and plants to a heap of jumbled leaves, branches and crushed hollow logs. Abrupt water-level rises associated with the low pressure in the eye of the storm would have led, then as now, to extensive flooding. Lightning would start vast forest fires, and the plants would burn to produce charcoal.

After this devastation new plants would seed and grow. Quickly the forest would re-establish itself, only to be devastated again by the next storm years later. The forest floor was probably often metres deep in rotting vegetation (Figure 12) destined to become peat and, much later, coal.

Figure 12 An artist's impression of how a Carboniferous coal swamp might have looked.

Question 6

Use Figures 9 and 10 to help you match the following sediments in a coal-bearing sequence A–E with the likely environment of deposition (a)–(e).

A   Shale with freshwater shells
B   Siltstone
C   Coarse sandstone
D   Seatearth
E   Coal

(a) Peat accumulations on the swampy areas of a delta plain
(b) Distributary channels cutting through the delta plain
(c) Fossil soil beneath the swamp vegetation
(d) Deposits laid on delta plain in times of flooding
(e) Shallow lakes and lagoons on the delta plain

2.2.4 The physics and chemistry of coal formation

How does rotting plant debris become a hard seam of coal? As plant material grows and dies, growing season after growing season, layers of vegetation will form on top of each other. If the region is subsiding, those layers of vegetation will eventually become covered with sediment. The weight of this sediment causes an increase in pressure which squeezes water and air out of the spaces and compacts the vegetation. As further layers of sand, silt and clay are added, the process of **compaction** continues. Eventually the vegetation will form a dense sediment, interbedded within sands and clays, indistinguishable from coal.

The first stage in the chemistry of turning plant material into coal is one of biochemical decomposition. This involves the breakdown of the more soluble components, principally the cellulose, with a resultant enrichment of the waxy, more resistant leaf coatings, spores, pollen, fruit and algal remains. Decomposition results in the loss of **volatiles** (the gaseous material given off by coal when it is heated) — chiefly water, carbon dioxide and methane — leaving organic compounds rich in carbon.

The second phase starts when the plant deposits are buried beneath substantial amounts of mud, sand and silt. *Coalification* involves chemical changes imposed by increasing temperature and pressure over time. The degrees of change are known as levels of **rank** and indicate the maturity of the coal. The different rank stages are listed in Table 2, together with some of the parameters used to define them. Changes in rank are gradual and so the boundaries of the rank categories are somewhat arbitrary.

Table 2 Changes in the chemical composition of carbon-rich material with increasing rank

Rank	Description	Typical content of volatile matter (weight %)	Typical element analysis (weight %)		
			Carbon	Hydrogen	Oxygen
peat	decomposed fibrous material	>50	50–55	5–10	≈40
lignite (soft brown coal)	still contains woody material; hard blocky appearance	45	>60	5–10	30–35
sub-bituminous coal	dark material developing slight lustre	45	<70 – >80	5–10	10–25
bituminous coal	black, shiny and strongly banded appearance	35	85	5–10	<10
anthracite	metallic lustre; bands absent	10	90	5	5
graphite	hard, with a silvery lustre	<5	>95	<4	<5

The continuous variation of chemical composition through the rank series is shown in Figure 13. Low-rank coals contain a higher proportion of volatiles than high-rank coals. This is shown by the variation in oxygen and hydrogen content in Table 2. **Anthracites**, for example, usually contain less than 10% volatile matter.

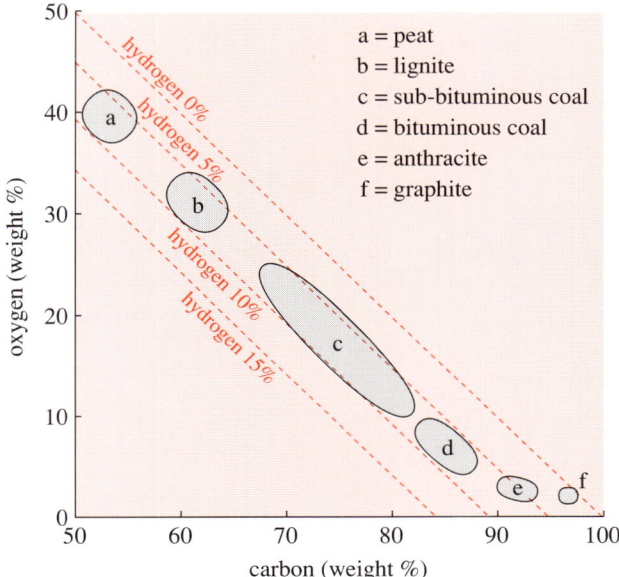

Figure 13 The relationship between carbon, oxygen and hydrogen contents for the six stages in the rank series.

The most important chemical change shown by increasing rank is the increase in carbon at the expense of oxygen. The proportion of hydrogen present remains relatively constant at 6–9% by weight over much of the rank series until about 90% carbon, where a significant reduction in the amount of hydrogen occurs.

The changes, summarized graphically in Figure 13, are due to the expulsion of water, carbon dioxide and methane. Only small amounts of methane (CH_4) are liberated during the early stages of coalification, but during the transition from **bituminous coal** to anthracite (particularly over the range 85–92% carbon) much hydrogen is expelled as methane and other hydrocarbons, whereas the emission of carbon dioxide declines.

- Would you expect the density of coal to increase or decrease with increase in rank?

- As rank increases, the porosity and water content of the coal decrease, so density increases.

The relative proportions of carbon and volatiles in coals affect their physical properties and their uses.

1. Coals rich in volatile matter (more than 30%) are easy to ignite and burn freely but with a smoky flame. Low-volatile coals are more difficult to ignite, but they burn with a smoke-free flame; they are natural smokeless fuels.

2. For industrial use, coal is usually heated to expel volatiles. The carbonized residue that remains after the volatile matter has been driven off in the absence of air is called **coke**. High-rank bituminous coals become partly fluid on heating and swell up to form a porous coke, especially valuable for the iron and steel industries. Coke makes a useful artificial smokeless fuel because it is free of volatiles.

3   The **calorific value** of coal (the amount of heat liberated under controlled conditions) generally increases with rank. Nevertheless, coals with a high volatile content (>30%) are usually burned in power stations for the ease with which they burn, even though they give out less heat than a higher ranked coal.

2.2.5 Impurities in coal

Coal is a complex organic material which contains impurities that originate both from the sedimentary environment in which it was deposited and from the chemical complexity of the source plant material.

Inorganic impurities within coal seams leave a residue or **ash** after combustion. This mineral matter either accumulated with the original plant material or was formed during coalification. The most common inorganic impurities within coal are clay minerals, which break down to give the common constituents of ash. The distribution of clay within the seam is important for practical reasons, since it determines the extent to which the coal needs to be cleaned before sale. The melting temperature of the ash and its tendency to corrode boiler surfaces are dependent on the composition of the mineral impurities. These are important considerations when assessing the suitability of coals as boiler fuels or for coking.

A second important group of impurities is the *carbonates*. During biochemical decomposition and the early stages of coalification, carbonate minerals are precipitated either as concretions (hard oval nodules up to tens of centimetres in size) or as infillings of veins within fissures in the coal seams. As we saw in Video Band 1: *The Great Iron and Steel Rollercoaster*, iron-rich concretions provided an important source of iron ore in Britain during the early stages of the Industrial Revolution. The fusibility of the ash is strongly influenced by the total carbonate content and by the proportions of calcium, magnesium and iron.

Nitrogen and sulphur are important minor constituents of plant protein, and are common impurities found in coal. The activity of microbes during coalification reduces nitrogen to ammonia (NH_4) and sulphur to hydrogen sulphide (H_2S). Small amounts of both of these can be trapped within the coal. Most sulphur is usually present as the mineral *pyrite* (FeS_2), in proportions varying from 0.5% to several per cent.

Another important impurity is sodium chloride (NaCl). When it is present in relatively high concentrations (over 1% chlorine content) the coal is virtually unusable industrially because of severe boiler corrosion.

Coal, like other sedimentary rocks, also contains a large number of trace elements in minute amounts. Some trace elements, including germanium, arsenic and uranium, are significantly enriched in coal. Although most of these end up in the ash after combustion, some trace elements may be released by burning coal and contribute to atmospheric pollution.

Now is a good time to view Video Band 9: *Energy Resources — Coal*. At this stage you need only concentrate on the first part, about the formation of coal (the first seven or eight minutes). We shall return to the later parts of this video in Section 3. It may help if you read the notes in the Video Box before viewing. After viewing, check your understanding of the first part of Video Band 9 by answering Question 7 about the formation of coal.

2 Fossil fuels and the carbon cycle

> **Video Band 9 Energy Resources—Coal**
>
> **Speakers**
>
> Howard Awbery   National Coal Board
> Ian Gass   The Open University
> Eric Skipsey   The Open University
>
> This programme was made in 1983. It has three main aims:
>
> 1. It identifies some of the geological complexities in coalfields and explains their origin by studying the way in which peat was deposited and transformed into coal.
> 2. It illustrates how the change from 'pick and shovel' mining methods to mechanization has made mining more inflexible, and as a consequence much more information is now needed about the geology of an area to be mined.
> 3. It illustrates the overall scale of coal workings and contrasts the coal deposits worked underground in Britain with the giant deposits worked by opencast methods in North America.
>
> The four mines visited are Babbington in Nottinghamshire and Butterwell in Northumberland (both now closed), Wabamun in Alberta, Canada and Sparwood in British Columbia, Canada.
>
> The following points made in the video are particularly important.
>
> - Variations of deposition within a delta system can result in variations in the thickness of coal seams and in the type of sediment that forms the roof and floor of the seam. Try to appreciate how these variations could affect the mechanized underground mining of coal.
> - At that time (1983) over 70% of Britain's coal production went to fuel power stations. Large power stations were then usually sited in rural locations in working coalfields, with no use made of the vast amounts of low-grade waste heat that power stations produce. The situation at Wabamun was similar.
> - The ease of extraction of near-surface opencast coal in Canada contrasts strongly with the problems of extracting coal from old British underground workings like Babbington. Even the mechanization of deep pits has not enabled them to compete economically with opencast workings. (In the ten years between the making of the video and the writing of this Course, Britain's deep pits were reduced in number from 70 to 17.)
> - Enough new reserves of high-quality coal in thick undisturbed seams have been identified in as yet unworked areas of Britain to enable us to meet all our electricity needs by burning coal if we so desired.
>
> Some new terms are used: *coal macerals* are the component parts of coal. *Vitrinite* or *bright coal* is the remains of wood tissue and bark, and *durain* or *dull coal* is composed principally of *inertinite* or material that is largely carbon and contains very little volatile matter.

Question 7

(a) What are the four main components of coal and how was each originally formed?

(b) According to the information in the video, what are the two of the most important impurities within a coal seam, and what happens to each when coal is burned?

(c) What is the sequence of rocks in a 'typical' coal cycle?

(d) How, according to the video, are washouts formed? What is their importance in mining terms?

(e) Describe the geographical environment of a coal swamp. When did swamps like this exist in Britain?

(f) Which geological process turns peat into coal?

2.3 The marine carbon cycle

So far in this Section we have only considered the terrestrial part of the carbon cycle. However we saw in Figure 5 that the oceans store much more carbon than the terrestrial system. How is this marine carbon fixed, and what are the main reasons for marine carbon fluxes?

Figure 14 shows the rates of natural carbon exchange within the marine system and between the marine system and the atmosphere, in billions of tonnes of carbon per year.

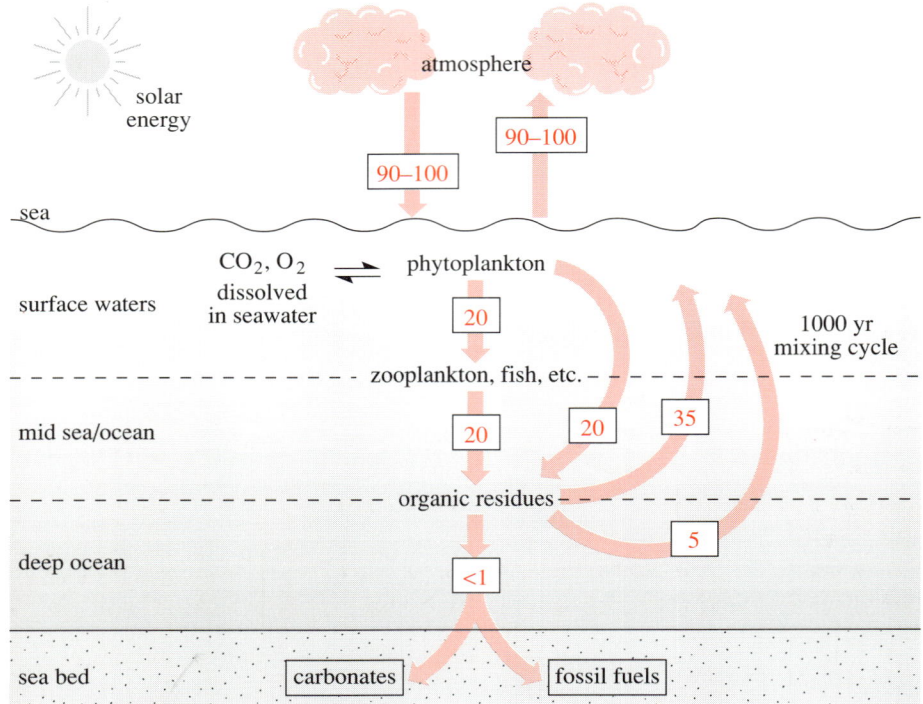

Figure 14 A schematic representation of the rate of exchange of carbon between stores in the marine environment and between the sea and the atmosphere. Rates are in tonnes of carbon $\times 10^9$ per year.

- Use the data presented in Figure 14 to state whether there is a net movement of carbon into or out of the atmosphere from the seas.

- Figure 14 shows that $90–100 \times 10^9$ tonnes per year of carbon flows from marine surface waters into the atmosphere, while roughly the same amount of carbon moves from the atmosphere into the seas. From this information there is some uncertainty whether slightly more carbon flows from sea into the atmosphere, or vice versa, or whether the system is in exact balance. In broad terms, though, there is an approximate balance in carbon flux each year from the atmosphere into the seas.

CO_2 is constantly being exchanged between the atmosphere and the upper levels of the oceans, by physical and by chemical processes. Worldwide, the flux of CO_2 into the oceans is broadly balanced by that back into the atmosphere. However, CO_2 is more soluble in cold water than it is in warm water, so the concentration of dissolved carbon is higher in cold polar waters than it is in warm tropical waters.

Marine phytoplankton (microscopic water-borne plant life) use dissolved CO_2 for photosynthesis. For that process they also need solar energy, so they must live in the sunlit upper parts of the oceans. Their photosynthesis releases oxygen which also dissolves in the seawater. Given light and sufficient nutrients, phytoplankton will bloom in the surface waters of oceans.

Cold, dense polar water sinks and flows under the influence of gravity along the ocean floor towards the Equator. It returns to the surface at *upwellings*, where surface waters are displaced by winds and currents. Upwellings occur in near-shore areas and in the open ocean. Wherever they occur, the nutrient-rich waters promote rapid growth of populations of phytoplankton.

Zooplankton (water-borne animal life, mostly microscopic) and higher marine organisms consume these phytoplankton. The waste products of zooplankton and larger organisms are large enough to sink through the water column, transferring carbon from the upper few hundred metres towards the deep oceans. Normally, little of this organic matter gets a chance to reach the ocean floor. It provides food for filter feeders, and through them for predatory animals, and ultimately for the ubiquitous decomposers. All these organisms release CO_2 back into solution through respiration. As Figure 14 shows, the marine carbon cycle is almost in equilibrium: less than 10^9 tonnes per year of carbon are available for incorporation into marine sediments.

- ● Assuming that 10^9 tonnes per year of carbon were incorporated into marine carbonates and fossil fuels in proportion to their present-day tonnages, how long would it have taken to deposit the carbon found in the current global store of fossil fuels?

- ○ We know from Figure 5 that the global carbonate store is some 9×10^{16} tonnes, and that the global fossil fuel store is some 10^{13} tonnes. If carbon is deposited into each in proportion, the rate of deposition of carbon into fossil fuels is $(10^{13}/9 \times 10^{16}) \times 10^9$ tonnes per year, or 1.1×10^5 tonnes per year. At this rate, it would have taken almost one hundred million years (more accurately, 91 Ma) to deposit the carbon to form the world's fossil fuels. Of course, not all fossil fuel formed in the marine carbon cycle: some of it formed on land.

2.3.1 Preservation of marine organic matter

Just as special conditions are required to preserve carbon from the terrestrial carbon cycle, so special conditions exist under which marine organic matter is preserved. Normally, marine carbon is recycled through decomposers in the food chain. To be preserved, therefore, any organic matter reaching the sea floor must be protected from scavengers or bacteria that would consume and destroy it. This protection is brought about by several factors.

1. Most scavengers cannot live in anoxic conditions, so where the bottom waters are depleted in oxygen, organic carbon can be preserved. Anoxic conditions often occur where water circulation is poor. For example, the present-day Oslo Fjord has an anoxic bottom layer because a shallow lip of rock prevents water from the Skagerrak from circulating around the fjord.

2. Most plants and animals cannot live in waters in which there is a high concentration of salts. Land-locked seas in tropical or subtropical latitudes can evaporate so quickly that the seawater becomes oversaturated in salts. Relatively oversaturated water is denser than normal seawater so it sinks to form a layer immediately above the sea floor. Under these circumstances, life can survive in the surface layers, but nothing lives on the sea floor or in the waters immediately above it. Organic material that sinks into the salty, lower layers remains undisturbed on the sea bed. The Dead Sea is a modern example.

3. There is a relatively high rate of sedimentation of inorganic material. A rapid supply of sediment means that organic matter will be buried rapidly even though sea-bottom conditions may support life. Such circumstances can produce anoxic conditions in the new sediment before the organic material has had time to decompose. Of course, substantial influxes of non-organic sediment dilute the concentration of organic material.

The Kimmeridge shales of north-west Europe

In Jurassic times the geography of north-west Europe was very different from today, and the continental shelf of what was to become Europe was an ideal site for organic-rich sediment deposition.

Thanks to extensive plate movements and continental drift, Britain lay in subtropical latitudes, at 30–40° N. There was no North Atlantic Ocean; instead a major rift valley led northwards from the area that is now north-west Europe into a polar ocean north of Greenland and Scandinavia. Large rivers drained northwards from Europe across coastal plains that lay between Scotland and Scandinavia.

In late Jurassic times some 150 Ma ago (Figure 15), sea levels began rising worldwide, probably due to increased plate tectonic activity. Areas of previously shallow water became deeper. Much of Britain and the North Sea area, formerly the site of coastal plains and river deltas, flooded to become a land-locked shelf sea.

A shallow sea situated in subtropical latitudes with links into a polar ocean was an ideal place for phytoplankton to thrive. Other organisms also colonized the upper layers of the sea; some grazing through the rich algal blooms, some feeding on the grazers. In deeper, offshore areas (such as the channel between Norway and Greenland) water circulation was restricted. Sluggish movement and a constant rain of organic material together reduced levels of dissolved oxygen in the lower layers, and the sea bottom became anoxic. The sea floor would have been a lifeless place, dominated by decaying organic material and sediment settling slowly out of suspension.

The undisturbed, millimetre-scale layering in Kimmeridge Formation rocks provides evidence for a calm environment, and also the absence of bottom-dwelling animals since fine layers in the sediment would have been disrupted by any feeding or burrowing. The sedimentary rock that formed in this low-oxygen, organic-rich environment was set to become the source rock for northern Europe's richest oilfields.

Question 8

Look at Figure 15 and answer the following questions:

(a) Which of these existed in Upper Jurassic times: Bay of Biscay, North Atlantic Ocean, Iceland?

(b) Which of these countries *could* contain Kimmeridge Formation carbon-rich shales: Denmark, England, Norway, Portugal, Wales?

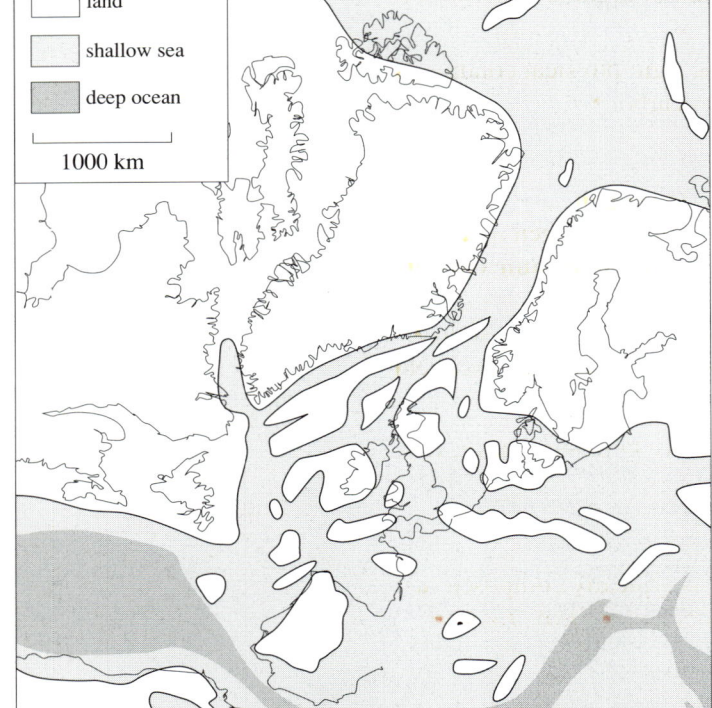

Figure 15 The geography of the Atlantic–European region some 150 Ma ago, when the Kimmeridge Formation was being deposited.

Sediment supply is greatest near to coasts and estuaries, where clay and sand is being swept from the land into the sea. However, these are not the best places to expect water circulation to be restricted. Wave and current action in seawater agitates and oxygenates it, so high-energy environments are not good sites for preserving organic material in an unoxidized state. Shallow waters, with their constant wave and current activity, may be very productive but little organic material will be preserved unoxidized. Where the water is deeper, the sea floor can lie undisturbed even during the fiercest storms.

But the main circumstances under which organic material is preserved is when the rate of supply of *organic material* is very great. Oxidation of part of the organic material de-oxygenates the surrounding water, allowing the rest of the organic material to be preserved free from oxidation. This is by far the most important factor; the other three are trivial by comparison.

There is no simple rule for exactly where in the marine environment organic-rich sediments will form. A rich supply of organic material is essential. Planktonic blooms within upwellings are therefore often important, as are anoxic sea-bed conditions. Carbon-rich marine rocks tend to develop under specific, rather special, sets of circumstances as our Kimmeridge case study shows.

Sediments that form under these conditions are usually fine-grained shales or mudrocks. They are frequently black because of the relatively high carbon and iron sulphide content. Such rocks are often called **oil shales**, and since they also form part of the suite of sedimentary rocks that are the source material for petroleum, oil shales are also known as **source rocks**.

For a source rock to be capable of yielding *significant amounts* of fossil fuels, two independent criteria must be met.

1. The source rock must contain sufficient amount of organic material for large quantities of hydrocarbons to form.
2. The source rock must have been subject to the right physical conditions for organic material to be converted into hydrocarbons.

2.3.2 Kerogen

Organic material in carbon-rich sediments is called **kerogen**, a word derived from the Greek for 'wax producer'. The concentration of kerogen in a potential source rock is usually expressed in terms of **total organic carbon** (the ratio of the mass of carbon in the rock to the mass of other rock-forming components). Rocks with less than 1% total organic carbon are not source rocks, whereas rocks with more than 10% total organic carbon are excellent source rocks.

Three different kinds of kerogen have been identified: kerogen types I, II and III. Each one is derived from a different source material, and each has a distinctive, hydrocarbon product. Table 3 highlights the major differences between the three kerogen types and shows how the chemistry and biological affinities of the three types vary. *Type II* kerogen is typically composed of the mixed remains of both phytoplankton and zooplankton. *Type I* is less

Table 3 The characteristics of the main types of kerogen

Kerogen type	Hydrogen–carbon ratio	Origin of organic material	Hydrocarbon that can be produced
type I kerogen	high (typically up to 1.6)	marine algae and phytoplankton	light, high-quality oil and some gas
type II kerogen	high	mixture of several types of marine plant and animal micro-organisms	main source of crude oil and some gas
type III kerogen	low (typically 0.8)	land-based source material, mainly plants	mainly gas with some oil, rich in wax
'dead' carbon	very low (typically much less than 0.1)	charred wood	no petroleum potential

> ### Kerogen in the Kimmeridge Formation
>
> The Kimmeridge Formation is an important source rock for North Sea oil. It has a total organic carbon content of 2–12% on average, rising to 31% where it outcrops in Yorkshire and up to 40% at Kimmeridge Bay in Dorset, England. Most of the organic carbon is type II kerogen with a H : C ratio varying from 0.9 to 1.2. Bacterially degraded marine algae, such as the phytoplankton genus *Tasmanites*, make up about 70% of total carbon; other marine microplankton and plant spores make up a further 20%; and the remaining 10% is derived from charred wood.

diverse. *Type III kerogen comes mainly from land plants*. Charred woody material, or charcoal, is almost pure carbon and is included in Table 3 for completeness. Most kerogen concentrations in source rocks are actually a mixture of each of these components.

2.3.3 Generating petroleum from source rocks

Under the right physical conditions, source rocks can produce **petroleum**. Petroleum is the general term for a mixture of organic compounds — liquids and gases — in which hydrocarbons with the chemical formula C_nH_{2n+2} (n can vary from 1 to over 40) are major components. There are seven main **petroleum fractions**, or groups. These are the *gases*, which include **methane** (CH_4), **ethane** (C_2H_6), **propane** (C_3H_8) and **butane** (C_4H_{10}); *gasoline* (C_5 to C_{10}; octane, for example, is C_8H_{18}); *kerosene* (C_{11} to C_{13}); *diesel oil* (C_{14} to C_{18}); *heavy gas oil* (C_{19} to C_{25}); *lubricating oil* (C_{26} to C_{40}) and *waxes* (over C_{40}). Naturally occurring petroleum is a complex mixture of various proportions of these fractions and other organic compounds.

The number of carbon atoms in a fraction determines its physical properties. Methane, ethane, propane and butane all have boiling points below 0 °C. The boiling points of the other fractions increase as the molecules get longer; for example, pentane (C_5H_{12}) boils at 36 °C, hexane (C_6H_{14}) at 69 °C and so on. Gasolines are therefore liquids and waxes are solids at atmospheric temperature and pressure. Melting points and densities follow the same trend, getting higher as the molecules increase in size.

Petroleum released naturally from kerogens will therefore be a combination of solids, liquids and gases. The gas form is known as **natural gas**, the liquid form is called **crude oil** and the solid form is known as **bitumen**.

Table 3 stresses another important difference between the three kerogen types, the petroleum fractions they are likely to generate. The chemistry of types I and II kerogen means that they are capable of producing *both crude oil and natural gas* under appropriate physical conditions. Type III kerogen usually yields natural gas, but very little crude oil. Pure carbon can yield neither, since it lacks the necessary hydrogen to form hydrocarbons.

Whether source rocks give rise to bitumen, crude oil or natural gas therefore depends in part on the chemistry of the kerogen in the source rock. Recognizing this, scientists call type III kerogen **gas prone** and types I and II kerogen **oil prone**. Remember that oil-prone kerogen can also yield gas and gas-prone kerogen can also yield oil under appropriate physical conditions.

When kerogen is heated under relatively anoxic conditions, chemical reactions related to that identified in Section 2.2.1 for the breakdown of cellulose enable it to release petroleum. The heating process is known as **maturation** (source rocks that can produce petroleum are termed *mature*). Since temperature increases with depth in the Earth, heating is naturally achieved by burial of the source rock. The actual temperature reached at a given depth depends on the rate of increase of temperature with depth, the **geothermal gradient**. Figure 16 shows the relative proportions of crude oil and gas formed from type II kerogen buried in a region where the geothermal gradient is about 35 °C km^{-1}.

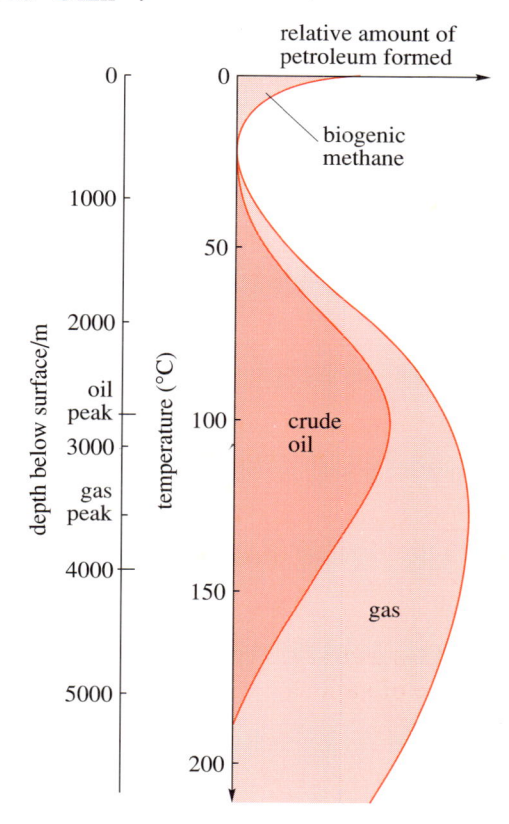

Figure 16 The relationship between depth of burial, temperature and the relative amount of crude oil and natural gas formed from type II kerogen in an area with a geothermal gradient of about 35 °C km^{-1}. Even after maturation some of the kerogen still remains unaltered as carbon residue.

- Look at Figure 16 and describe which type of petroleum will form below 50 °C, at 100 °C, and above 150 °C.

- Type II kerogen can yield both oil and gas. Some oil and some gas are generated at temperatures below 50 °C, but most of the kerogen remains unaltered. At 100 °C most kerogen is altered and the generation of petroleum is at its peak, of which over three-quarters is oil. At temperatures above 150 °C, gas is the principal petroleum formed.

Significant amounts of petroleum only begin to form at depths of between 1 km and 2 km, corresponding to temperatures above 50 °C (Figure 16). Peak oil generation occurs when source rocks have been buried to depths of around 3 km. At depths greater than about 4 km, temperatures are high enough for oil to become thermally unstable, and to break down or *crack* to gas.

Time is another important factor in the generation of petroleum. Similar source rocks subjected to the same temperature for different lengths of time would yield different amounts and possibly even different types of petroleum. This point is brought out in Figure 17. In real petroleum provinces, rocks are not held at constant temperatures for extended periods of time, as we shall see in Section 2.3.4.

Question 9

Assuming that the kerogen in the source rocks is suitably oil and gas prone, use Figure 17 to decide whether oil or gas will have been generated from the following source rocks:

(a) The source rocks were buried in a region with a low geothermal gradient and remained at a temperature of 50 °C for 30 Ma.

(b) The source rocks were buried in a region with a high geothermal gradient such that the temperature was 190 °C for 30 Ma.

(c) The source rocks were buried deeply in a region of high geothermal gradient where temperatures have reached 300 °C for 100 Ma.

Figure 18 shows the depth of burial of the top of the Kimmeridge Formation in the central and northern North Sea areas, indicating where oil could have been generated from Kimmeridge Formation source rocks. Compare the depth to the oil peak on Figure 16 with both the 3000 m contour and the sites of the major oilfields in this part of the North Sea on Figure 18; clearly, the major oilfields in the UK part of the North Sea are closely linked to the maturation of their Kimmeridge source rocks.

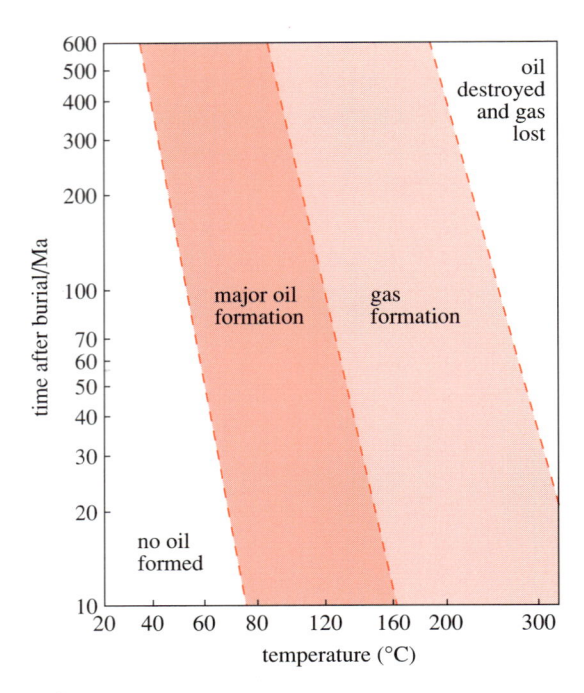

Figure 17 The relationship between temperature, time and petroleum generation. The time that any oil-prone source rock is held at any given temperature influences whether oil or gas forms. Note that the scales are not linear.

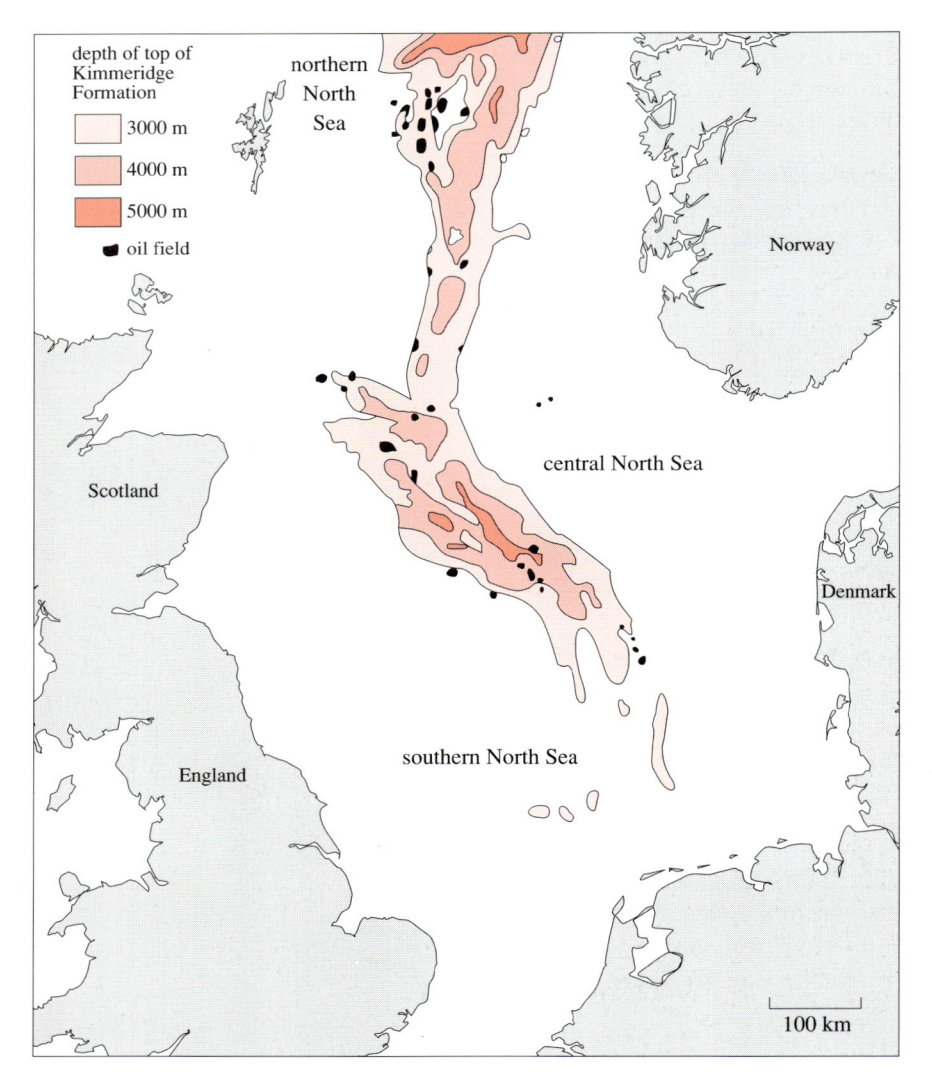

Figure 18 Map showing the depth to the top of the Kimmeridge Formation under the central and northern North Sea. Depths of burial are shown by 3000 m, 4000 m and 5000 m shaded areas. Some major oilfields are also shown, but not named.

2.3.4 Three examples

Let us now contrast petroleum generation in three areas where the geological histories have been quite different; the Paris basin in northern France, the Viking Graben in the northern North Sea and the Los Angeles basin in California. The source rock in each of these areas is of different age. Figure 19 shows the actual burial history of these three source rocks over time, and as you can see they cannot simply be related to the theoretical model of Figure 17.

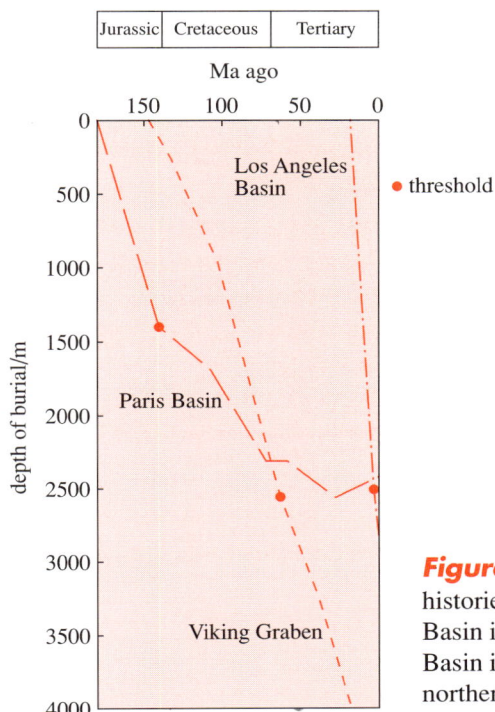

Figure 19 Reconstruction of burial histories of rocks from three basins; the Paris Basin in northern France, the Los Angeles Basin in USA and the Viking Graben in the northern North Sea.

The *threshold*, or start of petroleum generation, is dependent on depth, temperature and duration of burial. In the Paris Basin, the threshold was reached after 40 Ma when the early Jurassic source rocks were buried to a depth of 1400 m. The temperature was then 60 °C. The late Tertiary source rocks of the Los Angeles basin were buried to 2500 m in 10 Ma, reaching temperatures of 115 °C before the threshold was reached. In the Viking Graben, the threshold was reached after 85 Ma when the Kimmeridgian source rocks reached 80 °C.

It has been estimated that the petroleum in the Paris basin was generated over a period of 120 Ma. This contrasts sharply with the generation of petroleum in the Los Angeles basin which took less than 10 Ma. Because the geothermal gradient, the depth, and the duration of burial can all vary considerably from area to area, so too can the time taken for petroleum to be generated.

2.3.5 Impurities in petroleum

In detail, the chemistry of the source kerogen is almost as complex as the chemistry of the plants and animals from which it came, and the chemical composition of the resultant petroleum reflects this. Typical compositions of petroleum samples are shown in Table 4.

Table 4 Composition of typical petroleum samples

Element	Crude oil (weight %)	Natural gas (weight %)	Bitumen (weight %)
carbon	85.0	65.0–80.0	80.2
hydrogen	12.0	1.0–25.0	7.5
oxygen	less than 2.0	trace	7.6
nitrogen	less than 1.5	1.0–15.0	1.7
sulphur	up to 3.0	trace–0.2	3.0

> **The Victorian oil industry in the Scottish Midland Valley**
>
> For a short time in the nineteenth century, the Scottish Midland Valley was the world's most famous oil-producing area. The development of the Scottish oil shale industry began six years before the first oil well was drilled in America. A local man, James 'Paraffin' Young, found that the miners of this part of Scotland were lighting their homes by burning *cannel* (i.e. candle) *coal* from which oil oozed. Young discovered that the thick brown shales that occurred together with the cannel coal also yielded oil when heated. He established an oil-production plant at Bathgate to the west of Edinburgh in 1851, and an oil industry destined to last for 104 years was born.
>
> The oil shales of the Midland Valley were distilled in retorts in an air-free, steam-rich environment. When all the hydrocarbons had been distilled from the kerogen, the residue was almost entirely inorganic, made primarily from complex silicates from the original shale. The spent, reddened shales were dumped widely, forming artificial red-coloured hills some 50 m high, known locally as *bings* (Plate 38). The oxidized and carbon-poor chemistry of bings makes them barren places, devoid of flora even after being exposed to weathering for many years. Despite recent attempts to landscape and reclaim some of them, many remain as eyesores — harsh red reminders of the results of petroleum production without thought or care for environmental impact.
>
> The last Scottish oil shale mine closed in 1962, but recent estimates have shown that there are about 65 million tonnes of extractable oil shale in this part of Scotland. But put into perspective, this would produce only the same amount of oil as about four months' yield from the giant Brent oilfield in the North Sea.

Sulphur, an important minor constituent of both animal and plant tissue, is commonly found in petroleum in proportions varying from 0.3% to 8%, more significantly in bitumen and crude oil than in gas. (In fact, gas is almost sulphur free.) Oxygen is a significant impurity in bitumen, but is commonly only found in trace amounts in crude oil and gas. Nitrogen, however, is found only in trace amounts in bitumen and oil, but may be present in significant quantities, up to 15%, in natural gas.

The link between the traces of chemical impurities in source rocks and those in the petroleum they generate is often used by the oil industry to trace the source of oil in oilfields.

2.4 Gas from coal

Not all natural gas resources are associated with oil. Oil is produced when types I and II kerogens are heated sufficiently to crack, but type III kerogens ~~do not produce~~ [*usually produce very little*] oil at any temperature; they are not oil prone. Type III kerogen is woody material, typical of and usually produced by land plants — a description that could equally well be applied to coal.

It follows that if coal seams are heated sufficiently, their complex organic molecules may become broken down into simple hydrocarbons, especially methane, the main constituent of natural gas. As we shall see in detail in Section 4, much of the gas in the southern part of the North Sea is derived from coal, not from oil shales.

Now is a good time to watch Video Band 10: *Oil — Finds for the Future*. The first part (the first eleven minutes or so) is about petroleum reserve predictions. It is of interest, but not directly relevant to our studies here; we shall be returning to these concepts in Section 4. The middle parts (the next ten minutes) are most relevant to your reading of this Section, as they describe some experimental methods for producing petroleum, by *in situ* recovery of oil from oil shales, and by generating methane from carbon. When you have viewed Video Band 10, check your understanding by answering Questions 10 and 11.

2 Fossil fuels and the carbon cycle

> **Video Band 10 Oil—Finds for the Future**
>
> **Speakers**
>
> Geoff Brown The Open University
>
> L. F. Ivanhoe President, Novum Corporation of California
>
> M. King Hubbert
>
> This programme was made in 1983.
>
> The first part of this video discusses a technique developed by M. King Hubbert in the 1950s to estimate the lifetime of fossil fuel reserves based on discovery and production data. The conclusion Hubbert draws is that by the time this video was made in 1983 the US domestic oil industry had already passed its peak. One way of extending lifetimes is to recover a higher percentage of oil from existing fields. Various successful techniques had by then been developed, but the limit is reached when it takes more energy to extract a volume of oil than is contained in the oil itself.
>
> The programme then moves on to consider various methods of producing hydrocarbons artificially from source rocks. In certain Cretaceous sands in Athabasca, petroleum makes up 30% by volume of the rock. This is quarried and processed by heating in rotating kilns to produce oil. The resource contains an estimated 8.2×10^{10} tonnes of oil.
>
> Large-scale oil extraction in the future will depend on processing reserves that are found in deeper, relatively inaccessible strata. That problem has been addressed by Geokinetics Inc., who have used a simple fire-front technique to extract commercial quantities of oil from 'the world's first economic oil shale extraction plant', in Utah.
>
> Coal could hold the key to future supplies of liquid hydrocarbons. Gas has been produced from coal for over a century; up to the 1970s all gas supplied to British homes was produced this way. The principle is that complex organic molecules are broken down by heating into simpler hydrocarbons, using **catalysts** to enhance the reaction. The correct combinations of catalyst and reaction temperature are crucial if methane is to be obtained from coal efficiently.

Question 10

(a) Describe the Green River oil shales of mid-western USA. How did they originate?

(b) Describe the process used by Geokinetics Corporation to recover oil from *in situ* rocks.

Question 11

Write down the two chemical equations that express the reaction of carbon (graphite) with steam at (a) high temperature and (b) lower temperature. Which catalyst had, at the time of filming, been most effective in enabling methane to be produced from the lower temperature reaction?

2.5 Generating and concentrating carbon

Some important questions still remain unanswered. What is the origin of the carbon within the carbon cycle? What variations affect the carbon cycle over geological time? How much energy could the carbon within the natural system produce?

2.5.1 Volcanoes and the carbon cycle

Figure 5 showed that by far the greatest proportion of the global carbon store is locked into carbonate rocks. Over the several billion years of the Earth's history, carbon must have moved from the atmosphere into the oceans and thence into carbonates. How did the atmospheric carbon originate?

The Earth's atmosphere as a whole was derived mainly from gases brought to the surface from the Earth's interior. Most volcanic carbon comes from the steady degassing of lava flows rather than from volcanic vents. For example, the 1991 eruption of Mount Etna in Sicily released an estimated 13 million tonnes of carbon in the form of CO_2.

In the short term, volcanic sources release insignificant volumes of CO_2 compared with other fluxes of carbon, but over geological time, degassing of the Earth's interior can reasonably account for all the carbon in the natural surface system.

2.5.2 Productive environments, present and past

Which factors lead to a *concentration* of carbon in sedimentary rocks? Carbon is likely to be concentrated in rock sequences which themselves were deposited in regions of the world that naturally fix most carbon.

At the present time the world's forests grow mainly in tropical areas and across the temperate plains. Since the relative positions of the Earth's climatic belts are not thought to have changed significantly over time, substantial forests would almost certainly have grown in comparable geographical locations in the past.

Look at Figure 20, which shows a plate tectonic reconstruction of the positions of the continents as they were in late Carboniferous times, some 310 Ma ago. By comparing Figure 20 with Figure 4 it is possible to identify parts of continents that lay in the warm humid equatorial regions around the Equator and in temperate latitudes (60° N and S) in Carboniferous times.

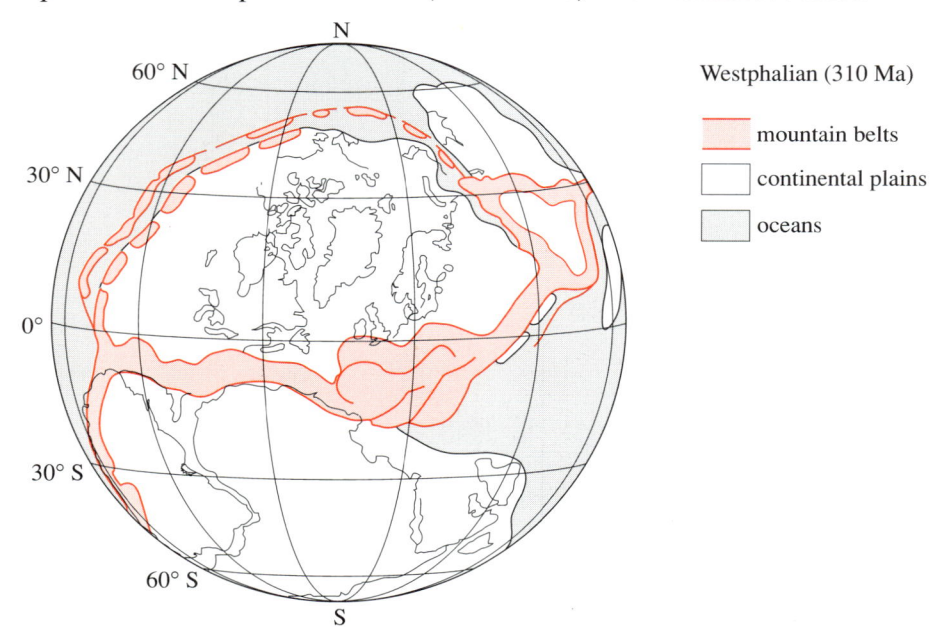

Figure 20 A reconstruction of the positions of the major continents in late Carboniferous times about 310 Ma ago. Most of the Carboniferous coalfields are found in areas that lay close to the Carboniferous Equator. (© 1993 The Geological Society)

Marine carbon concentrations are affected by changing mean sea levels with time. Mean sea level changes over geological time for many reasons. About

20 000 years ago, during the last Ice Age, sea level was about 150 m lower than it is today because a significant amount of global seawater was locked into extensive continental ice sheets. By contrast, 100 Ma ago in Cretaceous times, there were no significant ice sheets so there was more water in the oceans, and the volume of the ocean basins was reduced because of particularly active sea-floor spreading ridges. Mean sea level stood then some 300 m higher than it does today. Over geological time, mean sea levels have therefore varied over several hundreds of metres.

Changing mean sea level on this scale has a profound impact on shallow seas and coastal plains. Many of the present continental shelf areas are less than 200 m deep. A fall in sea level of 150 m would dry out many areas around the world, which would cease to produce marine organic carbon. Conversely, a rise of 300 m would flood vast areas of low-lying coastal land throughout the world, turning it into shallow sea and giving rise to much more extensive marine carbon 'factories'.

Times of high global sea level, such as happened in Ordovician–Silurian times (450–400 Ma) and Jurassic–Cretaceous times (200–50 Ma), are characterized by extensive shallow seas worldwide. Rock sequences of these ages often contain high amounts of carbon in limestones and in organic-carbon rich rocks; the latter form good petroleum source rocks.

2.5.3 Preserving carbon-rich rock sequences

The next point in this section is a brief but important one: what leads to the preservation of rock sequences that contain fossil fuels? Geological conditions have to exist that allow the preservation of the organic material as well as the entire rock sequence.

Falls in mean sea level would naturally expose coal seams and petroleum source rocks to surface erosion. For rock sequences to be preserved they must be covered by a steady supply of younger sediments, despite changes in sea level. This is brought about by *subsidence* of parts of the Earth's crust, itself often caused by stretching associated with plate tectonic processes (Block 1, Section 3.6).

Basins formed by crustal stretching are not only the best sites to preserve thick sedimentary rock sequences but are also the best places for petroleum to be generated from appropriate kerogens by burial. Seekers of fossil fuels concentrate their efforts on the sedimentary basins of the world, as we shall see in Section 4.

2.5.4 The fossil fuel bank

The fossil fuel 'energy bank' that has accumulated over the past few hundred million years in particular (the period of major coal and petroleum formation) has been estimated to contain 10^{23} J of energy. It is being added to by preservation of marine and land organisms at the rate of 10^{17}–10^{18} J every year. This latter figure represents roughly one-thousandth of current world energy demand.

Question 12

Our energy inheritance stored in the fossil fuel 'bank' is about 10^{23} joules. In 1992, world energy consumption from all sources was about 4×10^{20} joules. Assuming world energy production comes only from fossil fuels (almost correct), how long can fossil fuels be burned *at this rate* before the energy 'bank' would be empty, and our entire stock of fossil fuels burned?

However, this finite 'bank' only holds about 0.04% of the total fossilized organic material. Most organic material occurs in sediments and sedimentary rocks as finely dispersed animal and plant debris which will never be commercially exploitable. The *average* total organic carbon of sedimentary rocks is about 0.4%.

Despite their apparently wide distribution, fossil fuel fields represent a very restricted number of locations in space and time where this average has been greatly exceeded. We now move on to consider how these locations are discovered, and how their resource can be extracted for the use of humans.

2.6 Summary of Section 2

1. The world contains an estimated 9×10^{16} tonnes of carbon. Most of this is locked within carbonate rocks. An estimated 10^{13} tonnes of carbon is contained within fossil fuels.

2. Green plants use solar radiation to build carbohydrates and plant tissue from carbon dioxide and water in the atmosphere, in a process known as photosynthesis. Photosynthesis releases oxygen into the atmosphere. When they respire, organisms use oxygen from the atmosphere to generate energy from food, releasing carbon dioxide and water vapour back into the atmosphere. The respiration reaction is the reverse of photosynthesis.

3. When plants and animals die, the organic material in their bodies is normally oxidized. If, however, organic matter decays in an environment where oxygen supply is limited, organic compounds relatively enriched in carbon and/or hydrocarbons are produced.

4. Over millions of years, the Earth's natural carbon cycle has operated to create a balance between the production and use of atmospheric gases. Solar energy supply, atmospheric composition and plant growth are broadly in dynamic equilibrium.

5. Peat is a forerunner of coal. To be preserved as peat, plant growth must exceed decay and oxygen-poor conditions must become established. Swamps and bogs where the water table is at or above the surface provide an ideal environment for peat preservation. As peat swamps become loaded by sediment, water and air is squeezed out of the spaces in the lower layers and the peat compacts. Eventually coal layers may be formed, interbedded between layers of sand and clay.

6. Coal represents the fossilized remains of former land plants. Within its microscopic structure, fragments of bark, wood, spores, and resin can all be found. Land plant fossils are commonly associated with coal seams, and no pre-Silurian (i.e. pre-land plants) coals are known.

7. Coal forms in low-lying estuarine and coastal regions, and is preserved where there is overall subsidence of the region and there are adequate supplies of sediments to bury the peat.

8. With continued burial, volatiles are lost and both oxygen and hydrogen decrease relative to carbon. Pressure and temperature together change low-rank coals into high-rank coals. The calorific value of coal generally increases with increasing rank.

9. Concentrations of marine microplankton occur in the upper sunlit layers of the oceans where upwelling currents bring adequate nutrients. These form the basis of the marine food chain. There is only a build-up of carbon within marine sediments where there is an adequate supply of

organic material and where physical conditions are right for its preservation.

10. To yield oil or gas, the sediments must contain sufficient organic material compared with the total rock mass, and this carbon-rich sediment must be heated to temperatures of 100–200 °C to convert its kerogen into hydrocarbons. This heating is achieved by burying source rocks under 2–5 km of younger sediment. The length of time the source rock has been buried affects the amount of hydrocarbons generated.

11. Whether oil or gas is produced depends primarily on the type of organic material present in the source rock, the maximum temperature the source rock reached and the time it remained at that temperature. Type III kerogen, effectively the same material as coal, sources gas when buried deeply enough. Types I and II kerogen usually source first oil, then gas with increasing burial and/or time.

12. Crude oil can be obtained artificially, from oil sand, by underground distillation from oil shales, or from hydrocarbon residues.

13. Natural carbon fluxes have not been constant over geological time. In particular, rising sea levels lead to increased areas of shallow seas, and to an increase in marine carbon fixing. Two periods of high global sea level in the geological past, Ordovician–Silurian times and Jurassic–Cretaceous times, contain higher than usual concentrations of marine carbon-rich rocks.

14. Most carbon-rich rocks are found in sedimentary basins; but most organic material within sedimentary basins is finely dispersed and will never be commercially exploitable.

Question 13

Of the eight statements (a)–(h), six are correct and two are incorrect. Identify the two incorrect statements and say how the corrected statement should read.

(a) There is approximately 9000 times as much carbon locked into carbonate rocks as there is in fossil fuels.

(b) Marine and land plants fix carbon through photosynthesis, and both they and other organisms release it through respiration. The net result is that in the short term the natural carbon cycle is roughly in dynamic equilibrium.

(c) Organic material is normally recycled. Generally speaking, it is only under oxygen-depleted conditions that organic material is preserved and can form fossil fuels.

(d) Evidence from wood, spores and resin within coal, and external geological evidence like the age of the oldest true coal, shows that coal must have formed from land plant material.

(e) Oil source rocks form where there are still, anoxic or saline bottom conditions, high organic productivity in the upper marine layers and a steady rain of organic sediment.

(f) To yield significant quantities of oil, source rocks must have a relatively high total organic carbon and be buried to depths where the temperature can be held steady at 115 °C.

(g) Originally, carbon came from inside the Earth, and was released through volcanic activity.

(h) Most of the organic carbon preserved in sedimentary basins can be accessed for use as fuel.

Question 14

Figure 21 illustrates the relationship between carbon and hydrogen contents over the coal rank series. Mark the boundaries of the six rank stages and describe the physical characteristics of each stage.

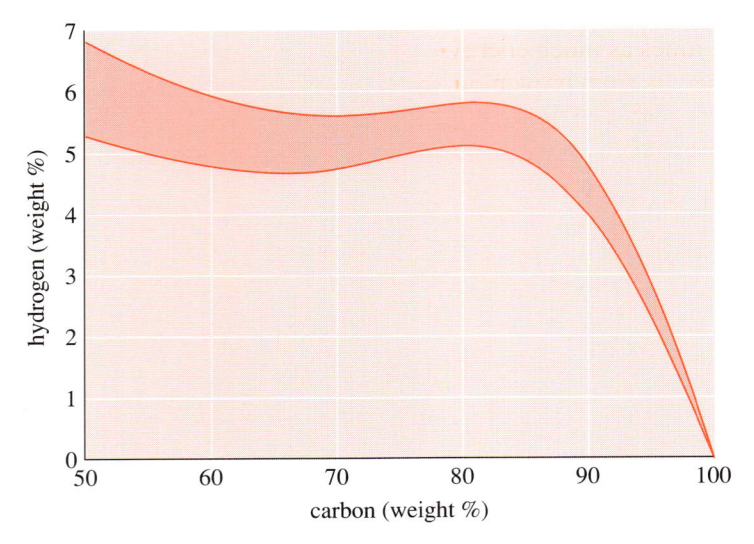

Figure 21 The relationship between carbon content and hydrogen content over the coal rank series.

3 FINDING AND EXTRACTING COAL

Coal is often regarded as the principal fossil fuel. Humans have been burning coal for thousands of years, and there is almost three times as much energy available from the world's proven coal reserves as there is from proven oil and gas reserves taken together. It is still the main fuel used to generate electricity; nearly half of world electricity generation comes from coal-fired power stations.

This section considers the techniques used in coal exploration, how coal is produced from surface and deep mines, and the environmental implications of coal mining. The section ends with an assessment of current world coal reserves, and consideration of the effect that changes in economic and political factors have on coal reserves.

3.1 Exploring for coal

Almost certainly, early humans discovered the fact that coal burned by accident, after a lightning strike or a forest fire left some rocks burning. Having discovered 'burning rock', they would have found it easy to trace the distinctive black colour and lightness of coal along an outcrop or to locate it elsewhere.

3.1.1 Mapping coalfields

Coalfields can be divided into two categories: **exposed coalfields** where the coal-bearing rocks outcrop at the surface, and **concealed coalfields** where they are hidden beneath younger rocks (Figure 22). Historically, mining activity began in the exposed areas of coalfields and then gradually extended into the deeper, concealed parts as the shallower seams became exhausted. The exposed coalfields can be defined with considerable precision by surface geological investigations; indeed geologists travelling on foot recording data represent the cheapest exploration 'tool' available to the coal industry.

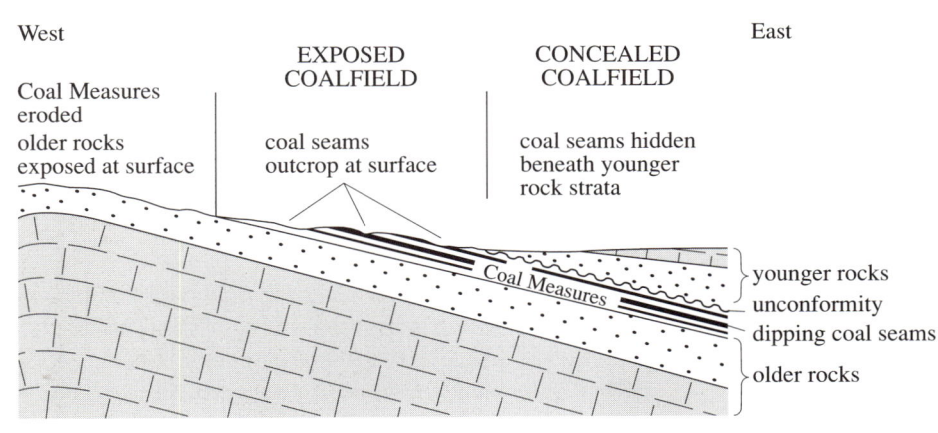

Figure 22 A generalized cross-section showing the distinction between exposed and concealed coalfields.

In Britain, the sites of coal at outcrop have been known for a millennium, and well-surveyed geological maps have been available for over a century. In fact all the exposed coalfields were mapped at a scale of 1:10 560 (six inches to one mile) by surveyors from the Geological Survey by 1925. In populated countries the chance of a surveying geologist finding new coal occurrences is negligible, but in remote areas of the world such finds are still possible, perhaps within dense unexplored jungle or under polar icecaps. Such areas are difficult to access, so surface data acquired from satellites or aeroplanes are used to aid the geological interpretation.

Drilling techniques

Drilling is the principal exploration method available to the coal geologist in both new and established areas. Boreholes may be drilled from the surface or from existing underground workings. Drilling is expensive, so over recent years much scientific and technological effort has gone into making drilling more sophisticated and precise.

The basis of all drilling is a *drilling rig.** Above ground, the most visible part is the tall *derrick*, which is essentially a crane. However, the drilling rods are stored and the drive motors located here (Figure 23). Most rigs use a rotary drilling technique developed in the USA in the 1920s. Below ground, the central tool is a steel, tungsten carbide or diamond-studded cutting *bit* that is rotated to cut a cylindrical hole through the rock sequence. Power is transmitted to the drilling bit through a series of steel pipes, known as the *drill string*. The drill string is attached to the bit at one end and to a rotating unit driven by powerful motors on the rig. As the drill string rotates, the bit turns at 75–250 revolutions per minute.

There are two designs of bit; a solid bit that is used for much of the period of drilling and a hollow, doughnut-shaped bit attached to a hollow *core barrel* which is used when samples need to be taken. The solid bit (Figure 24) grinds away the rock, reducing it either to chippings or to rock paste, since the bit is lubricated with water or some other drilling fluid. When it is in use, drilling can be continuous until the bit is worn out (many hundreds of metres on average) but no complete samples of the strata can be taken. The only indications of the type of rocks through which the drill bit is passing comes from the *chippings* or *cuttings* that are brought to the surface by the drill lubricating fluid (Block 3, Section 3.8.4).

Figure 24 A solid rotary drill bit.

If substantial rock samples are required, as is the case when a coal seam is penetrated, the solid bit is replaced by a barrel which effectively drills out a cylinder of rock, called the **core**. Coring long sequences of strata is a slow procedure. Either the whole length of the drill string or an inner core barrel has to be lifted out from the bottom of the drill hole at frequent intervals, the core removed carefully from the barrel, and the drill string or barrel replaced and relocated at the bottom of the hole. If the hole is deep this process can be very slow and costly. Because of this, techniques have been developed to investigate the nature of the rock sequence without going to the expense of coring all of it.

The secret of good core recovery is the modern core barrel. In the simple core barrel used in the 1930s the core revolved inside the drilling tubes as drilling proceeded. This frequently resulted in fragile or friable cores breaking into fragments, which fell back into the hole when the drill string was removed and were then pulverized as drilling continued.

The modern drill tube contains an inner core barrel which holds the core still as the drill revolves. The core is preserved whole and not broken during drilling. A further refinement enables the core barrel to be raised by cable inside the drill rods to the surface when full without all the rods having to be extracted.

To be of value, cores have to be carefully examined and the geological data recorded — a procedure known as *core logging*. The target coal seam and any others intersected are usually cored and logged, because precisely located samples are required for chemical analysis.

Figure 23 A land drilling rig. Cables raise and lower the drill string from the top of the derrick (1). The string is driven by a square-shaped rod (2) which itself is turned by a rotary table on the rig (3). The drill string is assembled from screw-in lengths which can be stored either vertically in or beside the derrick, or horizontally in a rack (4).

* There is no need to learn these terms in italics, they are included only for information and completeness.

3.1.2 Drilling

The only certain way of locating seams in concealed coalfields is by drilling boreholes (see Box on 'Drilling techniques'). Through drilling, a solid sample of the rock sequence at depth can be obtained, generally containing the full thickness of the coal seam or seams and often the rocks that lie immediately above and below the coals as well. It is then easy to see how thick the unit of interest is, how deep it lies below the surface and which rock types surround it. Coal samples are usually removed for chemical analysis, to measure its carbon, energy, sulphur and ash content.

3.1.3 Borehole logging

Geophysical techniques that record the nature of the strata penetrated in boreholes without coring allow faster drilling methods to be used for most of the strata. A string of electronic instruments is lowered by cable down the borehole, usually after drilling has been completed or whenever the string is lifted to replace a solid bit. Because the instruments are lowered by cable, rather than being attached to the drill string, the process is called *wireline logging*.

The electronic instruments are contained within a tubular logging device. Various tools are available to measure several physical properties of the rocks surrounding the borehole, but among the most useful tools are the two that record (a) the level of natural gamma radiation and (b) the rock density.

The natural gamma log

We saw in Block 1, Section 3.8, that some naturally occurring elements have radioactive isotopes. The two most common radioactive isotopes in rocks are of potassium and rubidium. In addition, radioactive isotopes of uranium and thorium are often associated with fossil-rich horizons. Minerals with high contents of any of these elements emit gamma radiation which can be detected with suitable instruments. Potassium is by far the most common, and it is present as part of the atomic structure of feldspars and some clay minerals.

Measurement of gamma-ray emission therefore provides an effective evaluation of the potassium content, and by inference of the feldspar and clay content of sedimentary rocks at any given depth in the borehole. The gamma log can be used to distinguish shales from siltstones, sandstones or limestones which usually have little to no potassium content and hence very low gamma-ray signatures.

Marine bands frequently show high radioactivity owing to gamma-ray emission from traces of radioactive uranium and thorium that are selectively concentrated in the shells of marine organisms. Shales rich in organic material often show very high radioactivity. Marine bands can often be identified by their extremely high gamma-ray characteristics from borehole logs alone. Coals have low gamma-ray characteristics since they rarely contain significant amounts of radioactive isotopes.

Rock density

Coal has a density of less than 1400 kg m^{-3} whereas the surrounding sediments often have densities of over 2000 kg m^{-3}. Density differences therefore provide a good method of distinguishing coal from other rocks without the need to examine cores.

Rock density is measured by exposing the walls of the borehole to gamma radiation emitted from an artificial radioactive source within the logging device. The amount by which this gamma radiation is scattered and re-emitted by the rock depends directly on the number of electrons per unit volume within the rock, which in turn broadly depends on the density of the rock. By using this specialized tool to induce gamma radiation, electron density and hence approximate rock density can be measured.

Figure 25 shows part of a real suite of logs obtained in a British coal sequence. Two logs are shown, the natural gamma log (calibrated in a specific unit defined by the American Petroleum Institute, API) and the density log (calibrated in grams per cubic centimetre, or 10^3 kg m^{-3}).

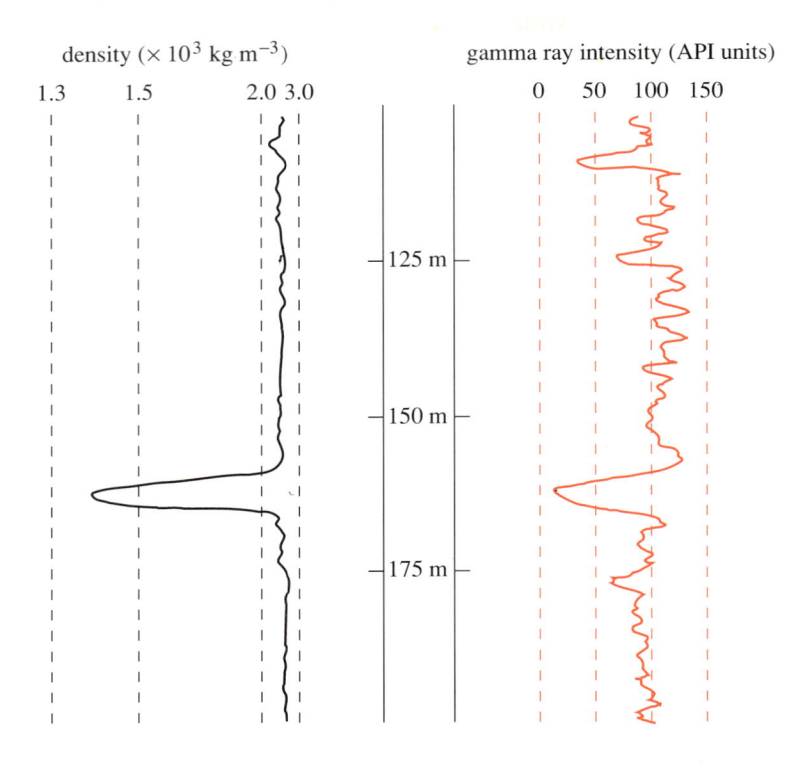

Figure 25 Typical natural gamma and density logs through coal-bearing strata. Natural gamma radiation is measured in comparison with an American Petroleum Institute (API) standard, and density is measured in grams per cubic centimetre. The central column indicates downhole depth in metres.

Question 15

From the gamma and density logs in Figure 25, decide whether coal is present in this borehole at depths of (a) 125 m and (b) 163 m.

Interpreting individual wireline logs is not easy, but cross-checks in the form of other logs are usually available to the logging geologist. If samples are not required for analysis, the technique is almost as effective and very much cheaper than continuous coring. In 1993 the cost of drilling 800 m with a solid bit and then coring to 1000 m was in the region of £100 000.

The thickness and quality of a coal seam in any area can be determined by drilling a grid of appropriately spaced boreholes. Initial investigations will typically be made using a grid of one borehole every 2–4 km^2, rising to five boreholes per square kilometre during production. Boreholes are not an effective means of identifying faults within a coalfield, because these are highly localized disruptions of the rock layers.

3.2 Winning coal in former times

Coal has been worked in Britain since at least the thirteenth century, and probably for much longer. Documents dating from 1228 tell of coal being shipped from the Tyne collieries to London and elsewhere. How did our ancestors find and work coal?

3.2.1 Early approaches

At first, coal was dug from strata exposed at the surface. Coal could be picked up from the ground, or from beaches below cliffs where coal outcropped and which were constantly being eroded by the sea. Mines were simply trenches following the coal seam, known as **drift mines**. Compared with modern working methods, the amount of coal that could be extracted from an open drift was small, even when wooden props were used to stop the roof from collapsing.

In the main the problems of ventilation and drainage limited the size of drift mines. Once worked, that section of outcrop was effectively exhausted even though the miners knew that more coal lay underground, beyond the limits of their mining and drainage techniques.

As demand for coal increased, new methods were developed and employed, and the **bell pit** became common. Miners would dig a vertical shaft down to the coal seam some distance from the coal outcrop. Once the underground seam was reached, miners would work their way outwards in all directions from the bottom of the shaft. Using bell pits, miners had access to artificial 'outcrops' of coal and could work the same seam from many access points. Coal was removed in baskets carried up the shaft by ladder. When the bell pit became unsafe, the shaft was simply abandoned and a new one started nearby. Bell-pit shafts were rarely more than 10 m deep and were usually 20–30 m apart (Figure 26).

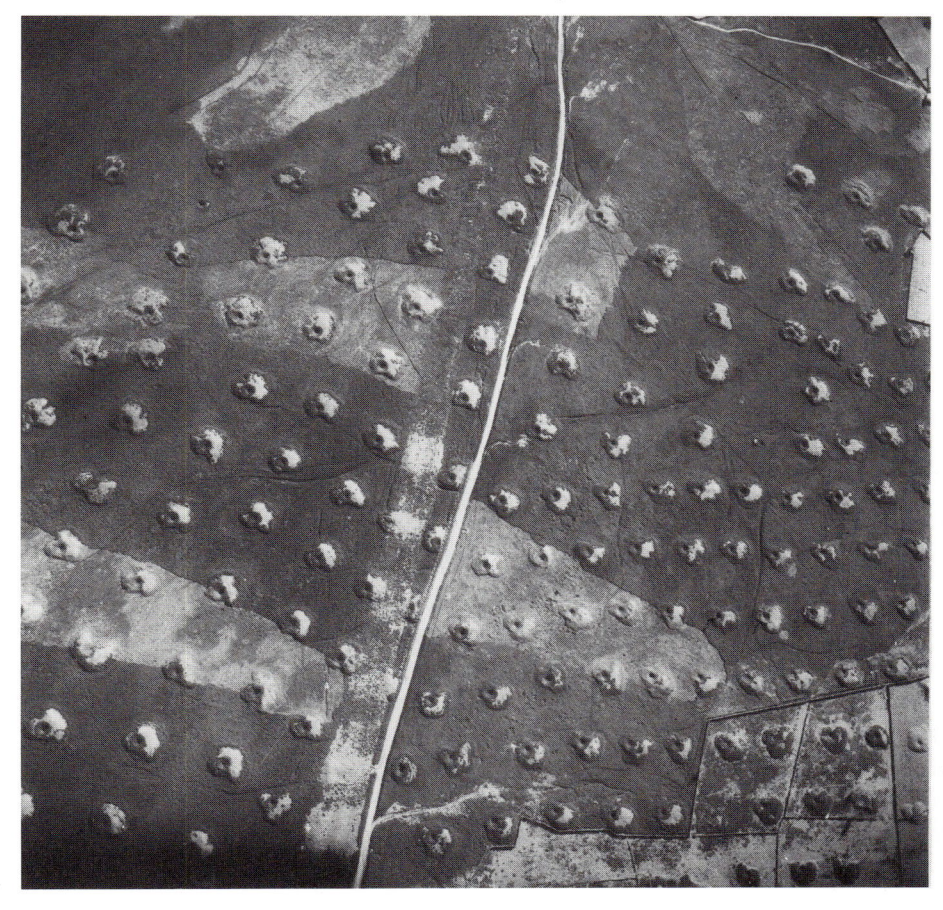

Figure 26 An aerial view of abandoned and collapsed bell pits working thin Jurassic coals on Rudland Rigg, North Yorkshire.

3.2.2 Pillar-and-stall working

Where seams lay deeper than 10 m, the repeated sinking of shafts was time consuming and wasteful, so a mining technique known as *pillar-and-stall working* was adopted. *Headings* were driven horizontally into the coal from the base of one main shaft. Initially, an almost random array of headings was used where up to half the coal had to be left in place, so the value of a rectangular grid was soon recognized. The coal was then divided into large blocks by *roads* at right-angles to the headings, and extracted from *rooms* or *stalls*, leaving wide *pillars* of coal as a roof support. Usually, the road-heading network was first driven into the whole area to be worked (Figure 27a). Later, pillars of coal may be systematically removed from the furthest limits of the mine back towards the shaft, allowing the roof to collapse (Figure 27b).

Pillar-and-stall working was ideal for mines with less than about 300 m of rock overlying the coal, but when the soil and rock cover was greater than 300 m, the great weight of the overlying rock often crushed the pillars. Ventilation also became a problem at these depths. To overcome this at least partially, the system of **longwall mining** was developed in Shropshire during the late seventeenth century and was in widespread use after 1850. This technique involved a 'wall', usually about 30 m long, being cut into the coal seam in one continuous process, advancing away from the shaft (Figure 27c). Any stone available (for example, sandstone within the seam) was built into support walls 2–5 m wide running at right-angles to the working face. These walls and wooden props were used to support the roof after the coal was cut and to prevent crushing of the coal at the working face.

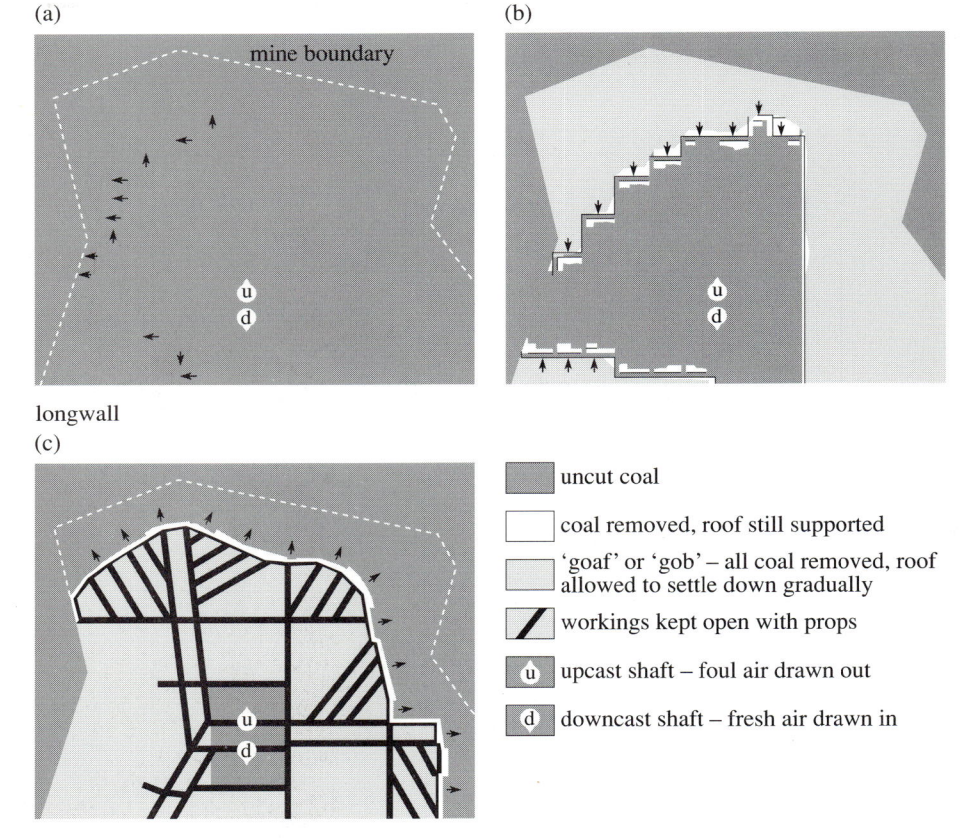

Figure 27 (a) Pillar-and-stall workings involved driving headings and roads in a right-angle grid throughout the whole area to be worked, leaving support pillars. (b) The roof was allowed to subside after the pillars had been removed. (c) Later, longwall workings became common, in which coal was removed as working advanced.

3.2.3 The problems of water and air, and the solutions

Once mining had advanced away from drifts and bell pits, the problems of actually cutting the coal became minor compared to the problems of exploration, drainage and ventilation.

Mines were limited in depth and distance away from costly shafts, locating coal was inefficient, and the only means of advance exploration was by digging trial shafts. The first recorded exploratory drilling, as opposed to shaft sinking, was in 1606 when Huntington Beaumont demonstrated his 'art to boore with iron rodds to try the deepnesse and thicknesse of the cole'. The first technique which allowed cores to be extracted was developed in 1804, and shortly after that steam power speeded up and reduced the cost of exploration.

As soon as shafts were sunk beneath the water table, flooding became a problem for the miners. Initially, long drainage channels were dug, or horse- or hand-operated buckets were used for drainage. Nevertheless the mines had to be abandoned when they flooded faster than they could be drained. Steam engines were introduced for drainage in 1712 and developed rapidly. After that, flooding itself posed few problems.

When mines were so effectively drained that coal could be worked a long way from the shaft, ventilation became important. Most early mines relied on the natural flow of air to provide ventilation. In bigger mines, forced ventilation had to be used, and **fire baskets** were developed for this purpose in the mid-seventeenth century. Fires were lit underground at the foot of an 'upcast' shaft — one of a pair of shafts close together. The upward movement of hot gases induced a partial vacuum in the mine, and a corresponding draught of fresh air into the mine through the nearby 'downcast' shaft. The flow of air through the mine was controlled by sets of trapdoors which closed off roadways. This system was effective and continued to be used until it was replaced by mechanical air pumps in the 1830s.

Life underground in the early 1800s

The early nineteenth century represented possibly the most dangerous and difficult times ever known for miners and their families. Gone were the small-scale productions of the bell pits of earlier times; yet to come were safety devices such as electrical or compressed air tools.

Both coal and miners were hauled up and down the shafts by rope. The men would put their legs through loops and cling on to the rope while they were being raised as much as 1800 ft (850 m). The boys reportedly found it 'fine fun', but it was extremely dangerous (Figure 28). Alarming episodes were recorded, like that of the young boy lowered 600 ft (183 m) down the shaft to start his shift. When the rope was raised the boy was still attached, his arms clasped around the rope, but he was fast asleep.

In the deep mines, explosions were commonplace. In Wallsend on Tyneside, one pit suffered nine explosions in the years between 1782 and 1820. In October 1820 a shaft was sunk to a lower seam, 900 ft (274 m) below the surface. So much methane gas was released while working this seam that miners lit the mine with it, simply by drilling a hole in the seam, inserting a tin pipe, and lighting the end. A pair of cannon were on hand underground to extinguish any fires too fierce to be put out by men with wet sacks. Just one year after the shaft had been sunk, 52 workers were killed in an explosion.

Figure 28 Mining in the early eighteenth century. Travelling up and down the shaft was a risky business.

3.2.4 The tools of the trade

Before the late nineteenth century, miners relied on picks, shovels and crowbars to win coal. Their strength lay in their versatility; miners working the coal by hand could cope with changes of thickness of the seam for whatever reason. Explosives were introduced in 1800 to drive roads and free coal.

Mechanized tools were developed with the coming of compressed air in the mid-nineteenth century. Toothed discs powered by compressed air were introduced in 1863, and rotary or continuously moving cutters became the norm. Electric conveyor belts which transferred the cut coal to roadways at the end of the face were introduced in the early twentieth century.

3.3 Modern coal production

Modern mines, in Britain at least, are much cleaner, safer and healthier places to work than those of a century ago. But modern mining is still changing. In 1940 all coal, and in 1980 over 90% of coal, came from underground methods. In 1993, over 20% of British coal was produced by surface (opencast) mining rather than underground methods, and the proportion is still increasing. In fact, the bulk of modern coal production worldwide is by opencast methods.

3.3.1 Surface mining

Producing coal from surface quarries is technically very similar to quarrying limestone and shale for cement manufacture. Like some limestones, coal seams lie buried under *overburden*, which may be sandstones, shales and limestones within the coal cycles themselves, or younger rock sequences including glacial material, or both.

Removing overburden is the main problem for opencast miners. We saw some of the techniques used for overburden removal in Block 2, Section 5 and Video Band 5: *Rocks for Roads*, particularly in the quarry at Dunbar. You will recall that at Dunbar a bridge conveyor system has been installed, which is able to move 1500 tonnes per hour of overburden and *interburden* (the waste material between limestones).

You may want to review the first part of Video Band 5 to refresh your memory about quarrying techniques before reading further.

Figure 29 shows schematically how opencast coal mining is organized. Topsoil is removed and stored or used immediately for land restoration. Shallow or soft overburden is removed by shovels and dump trucks but at deeper levels more powerful shovels have to be employed (Plate 37). Some harder layers of rock have to be broken using explosives. When the first coal is reached, seams are usually worked on each site by **bench mining** methods. The top surface of coal exposed on each bench is carefully cleaned to remove any adhering waste rock. The clean coal is then excavated by hydraulic digger and loaded into trucks. This careful exclusion of non-coal material, the *dirt* or *stone* impurities, enables consistently high-quality coal to be produced by opencast mining.

- Thinking back to Block 2, Video Band 5 and the Dunbar Quarry, what do you think controls the maximum depth to which coal seams can be worked economically using opencast methods?

- It cost money to employ workers and operate machinery to strip unprofitable overburden, so the maximum economic depth for opencasting depends on the value of produced coal compared with the cost of removing the waste material.

3 Finding and extracting coal

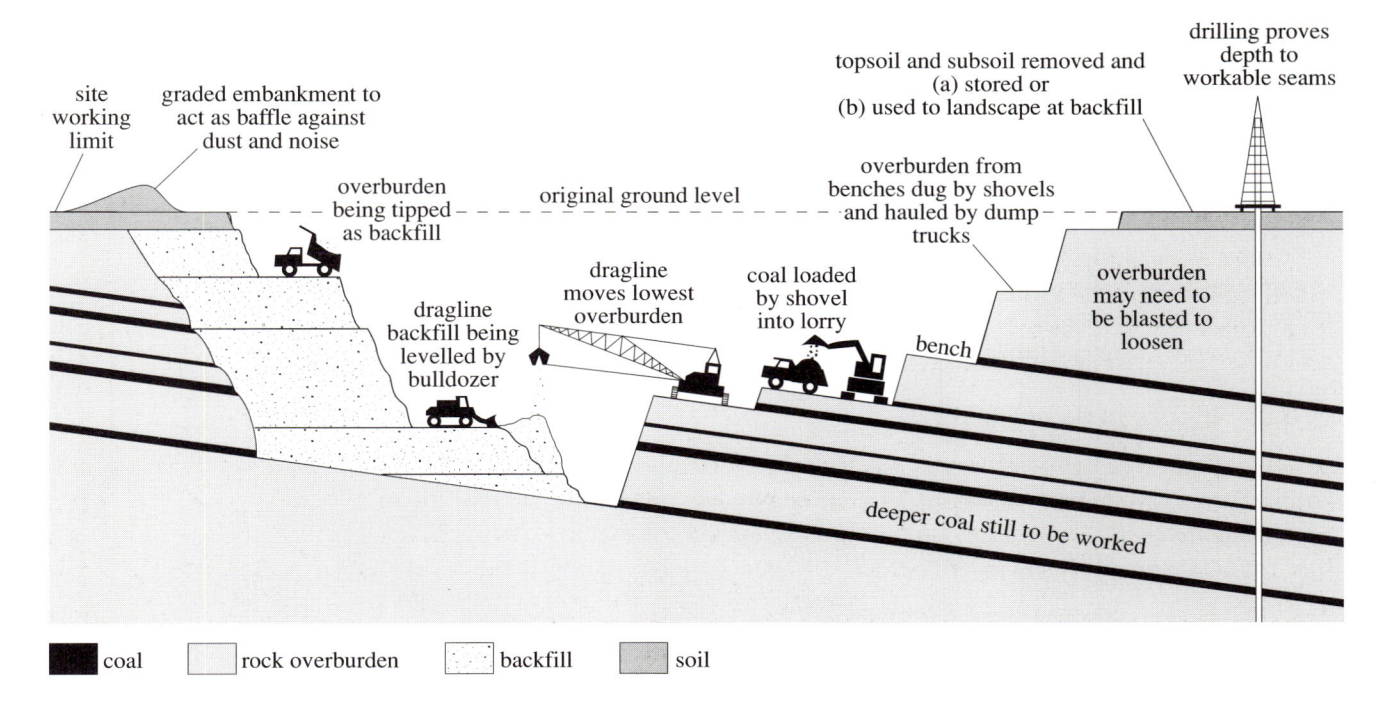

Figure 29 A diagrammatic section through the opencast mining process. Although unworked coal seams are shown to the left of the working area, in reality such shallow coal would have been worked long ago, or at an early stage of the opencast operations.

The main limitation on depth in opencast mining is the ratio of the amount of overburden to the total amount of workable coal. This was called the overburden ratio in Block 2 but within the coal industry it is more usually known as the **stripping ratio**. Stripping ratios can be calculated either in tonnes (as you saw at Dunbar in Video Band 5) or in thickness ratios (commonly used in coal operations). Mines working high-rank coal can operate higher stripping ratios than those working lower-rank coal; for example, opencasting provides most of the anthracite mined at present in the UK.

Surface mining benefits from economies of scale with the use of very large capacity earth-moving equipment for high productivity. The maximum economic stripping ratio has steadily increased over the years, as the productivity and life of the plant and equipment has improved. Economic stripping ratios are at present commonly around 20 : 1, rising to 35 : 1 in anthracite workings in south Wales.

In Britain, opencast sites are now located in areas of shallow coal where underground mining has ceased and includes districts where the old mining methods had extracted only a proportion of the coal. The average working depth in Britain in 1993 was 84 m. The deepest opencast operation, 150 m, was at Westfield in Scotland, made possible because of an abnormally large proportion of coal and coaly material. The ratio of total strata thickness to coal at Westfield is 3.2 to 1.

- What is the thinnest seam that can be worked economically using opencast methods?

- When bench methods are used, every seam can be worked, not just the thickest. Seams as thin as 0.1 m are worked where they form part of the overburden to a lower, profitable seam.

Tonne for tonne, surface mining is less costly and more flexible than underground mining, and recovers a higher proportion of the coal in the strata. Britain has 300×10^6 tonnes of potential opencast coal proved by borehole drilling programmes; at 1992 levels of output and using current technology this represents about 30–40 years of opencast coal. The typical 'life expectancy' of an opencast operation in Britain is around 5 years. Between 8 and 10 sites finish production each year, to be replaced by up to 10 new sites, although for example in 1994 the number of replacement sites did not exactly equal the number of decommissioned sites, leading to a slight reduction in the total number of sites.

Overseas, surface mining provides most of the coal from new coalfields. Some 20% of British coal is opencast, compared with a world average of 40%. The USA opencasts 50% of her coal; Australia, 70%; and Venezuela and Columbia close to 100%.

3.3.2 Underground mining

In areas where workable coal seams are buried to depths greater than the economic stripping ratio, underground mining methods must be employed. Modern underground mines rely on highly mechanized extraction techniques that were developed during the 1960s and 1970s.

The coal seams exploited in most British and Western European mines are typically about 1.5 m thick, but can vary from 0.5 m to 3 m. This coal is produced by longwall methods in almost all British and European underground mines. A modern coal face (Plate 39) is a production line between 50 m and 250 m in length, supported by hydraulic chocks and equipped with one or more cutters or *shearers*.* The shearer cuts a strip of coal and the face is then advanced by moving the hydraulic chocks forwards. As the face is advanced, the roof is allowed to fall into the cavity (*goaf*) formed when the coal is removed (Figure 30).

Tunnels or *gates* at either end of the face link it to the colliery's communication system, and are used to supply and ventilate the face and remove the coal.

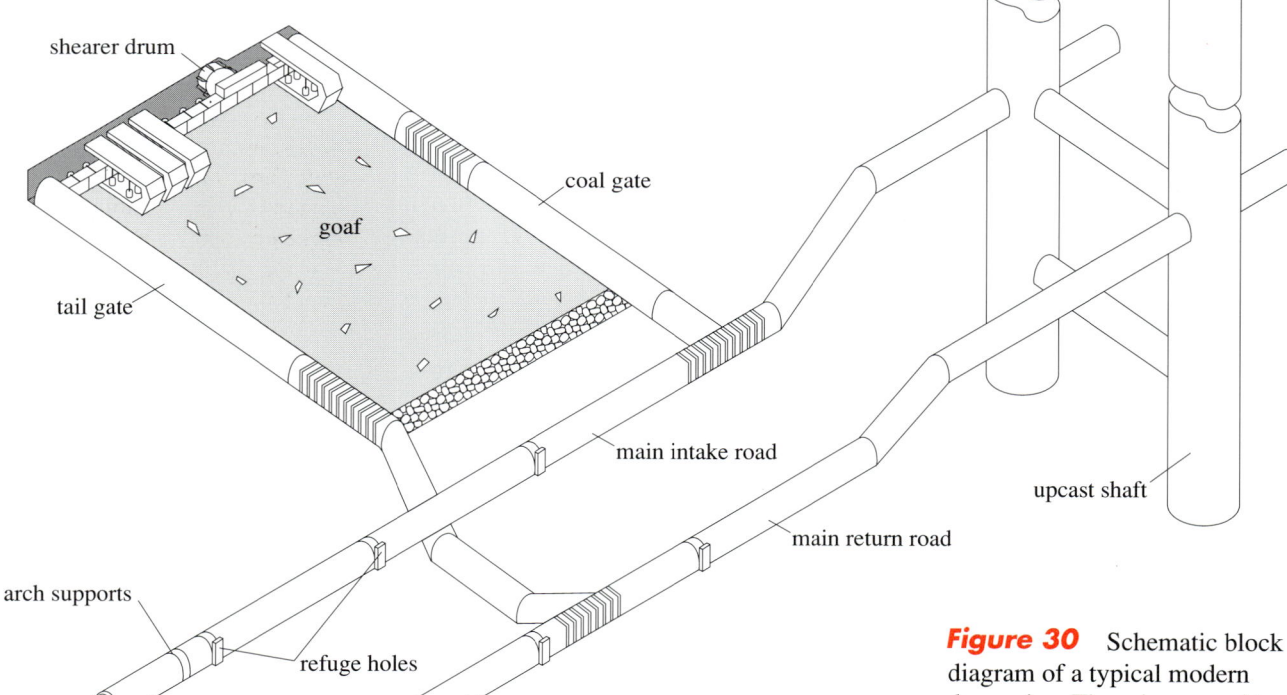

Figure 30 Schematic block diagram of a typical modern deep mine. There is no need to memorize these names as long as you understand the purpose of each of the components. (©1993 The Geological Society)

* The terms in italics in this Section are included for interest and information only — you do not need to memorize them.

3 Finding and extracting coal

In both new and established mines, roads have to be driven to new areas of coal working. A specialized tunnelling machine called a *roadheader* is used to do this. Cutting gates and roads is not a profitable operation in itself, but it does provide valuable geotechnical information to assist the mining operation. Like faces, roads and gates require supporting, which nowadays is mostly achieved by using concrete linings or steel arches with a pattern of roof and sidewall bolts.

The road network links the faces (most modern mines have several working at any one time) to the two shafts (Figure 30) that allow coal and stale air to be removed, and fresh air, workers and supplies to have access to the mine.

Figure 31 illustrates the two common types of underground layout used in British mines. At an **advance face**, the coal faces are advanced into a block of coal. The access gates at either end of the faces are also advanced to keep pace with the faces. At a **retreat face**, the access gates are first driven to the far boundary of the block of coal before the coal face is opened at their extremities. The face is then worked back towards the main roadways.

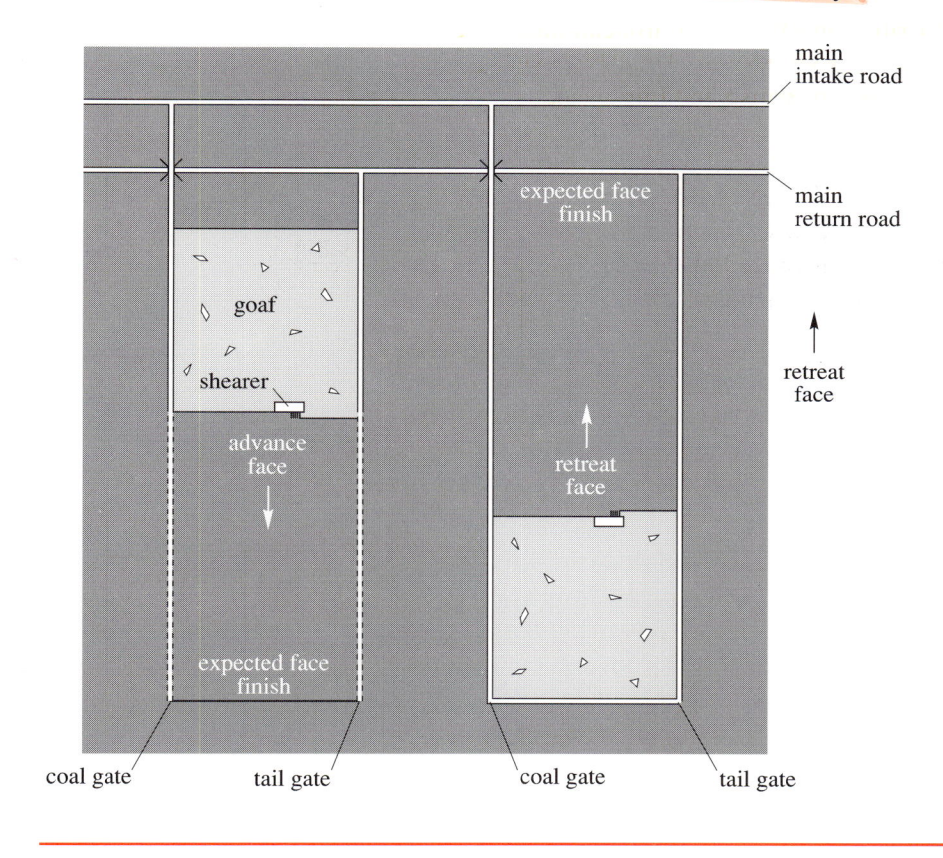

Figure 31 Two types of modern coal face layout. In the advance face, the coal face and gates advance into a block of coal. In the retreat face, the access gates are first driven to the boundary of the block of coal. The coal face is then opened at their extremity and worked back towards the road network. (©1993 The Geological Society)

Activity 3 Which factors interfere with production underground?

Plate 39 gives a valuable insight into the strengths and limitations of underground mining. Study the Plate carefully and use the information in it to answer the following questions.

(a) Which types of technical problem might lead to a temporary halt in coal face production?

(b) Which geological problems might interfere with coal face production?

It would seem sensible to try and extract all the coal in a seam but in practical terms this can never be achieved. Coal may have to be left at the top or bottom of the seam to provide better roof or floor conditions. Areas where the seam is disturbed geologically may not be workable, and support pillars may need to be left to prevent roof collapse and subsequent damage to underground roadways or surface buildings.

3.3.3 Geological problems in coal mines

A modern coal face is a very complex mining system representing a large investment in terms of capital, labour and planning. Modern coal-cutting systems are, in the main, inflexible and require uniform conditions to maximize output. What then are the effects of geological variations on such an inflexible mining system?

Geological factors are the primary controls governing the selection of working areas. There are two principal categories of geological conditions which affect mining operations. The first relates to the nature of the strata and lateral variations in rock type. The second concerns the **dip** of the strata and the presence of faults.

Variations in thickness and type of rock may occur both within the seam and in the strata forming its roof and floor. The thickness of a seam has a considerable bearing on the profitability of mining operations. Seams 1–2 m thick are particularly suited to mechanized longwall mining because the roof support equipment is most effective in this range. Seams less than a metre thick suffer from the law of diminishing returns since to obtain the same volume of output from a thin seam as a thick one, a greater area of extraction (and hence more rapid face advance) is required. The presence of shale layers within a seam results in reduced value, because the marketable value of coal is determined by its quality as measured by ash content. Such layers may also indicate the beginning of a split within the seam.

Of equal importance in achieving rapid face advance is the nature of the roof and floor strata. Shales and siltstones usually provide good working roofs for a coal face, but soft seatearths below the seam reduce quality because heavy equipment can gouge out several centimetres of floor material with the coal.

The presence of sandstones can result in operational problems on the coal face. The most serious problems arise where **channel-fill deposits** are encountered in the shales above a coal seam. These structures formed when erosive drainage channels on the original delta plain cut down into the underlying sediments and filled with coarse sand. If such channels cut down into the shales above a seam, they usually result in unstable roof conditions, and serious roof falls can occur. However, sometimes the channels cut down into the coal seam itself and, in extreme examples, may locally have eroded it away completely (Figure 32). Where these **washouts** occur they will bring a working face to a standstill.

Figure 32 Diagrammatic cross-section of a washout. A river channel has cut down into shale and peat layers shortly after their deposition, and subsequently become filled with sand. The continuity of the coal seam is disrupted, making underground mechanized working difficult.

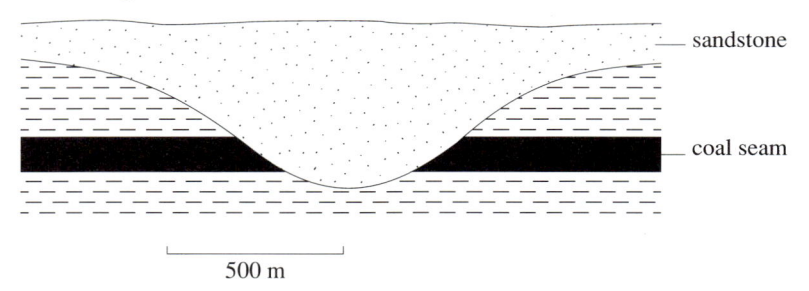

Whereas the sedimentary setting of a seam and its associated strata determines the profitability of the mining operations, the structural setting imposes physical constraints on the layout of mine workings. Faults form major problems for mechanized faces, as any displacement greater than seam thickness may bring production to a halt with consequent loss of output.

Firedamp

One of the best known hazards of underground mining is methane, or *firedamp* — the cause of so many explosions in underground workings in the past. Firedamp itself represents only one of several gases released from the seam when coal is worked.

Methane, carbon dioxide and water vapour become trapped in coal and its associated sediments during burial. When coal is mined underground, the trapped gases are released into the confined atmosphere of the pit. (Surface mining allows these gases to escape directly into the atmosphere.) Large quantities of carbon dioxide mix with nitrogen in the air to form a heavy, suffocating gas termed *chokedamp* or *blackdamp* by early miners. The existence of chokedamp demanded the establishment of an underground ventilation system; only by physically moving the heavy gas in the airflow could safe conditions become established.

Potentially more lethal was the mixture of methane and air that was also released. Firedamp is a mixture of 5–15% methane and 85–95% air which is highly inflammable, so the very act of freshening the air to remove chokedamp in deep mines led to the establishment of ideal conditions for a firedamp explosion.

Early steps to combat firedamp were crude, primarily because the miners of the time did not understand the nature of the problem. A pit canary would register both chokedamp and firedamp by suffocating and falling off its perch. It was obvious to the early miners that naked flames ignited firedamp, but the problem was to provide a light without a flame. At one stage, working in the luminescent light from putrefying fish was proposed; the faint glow and the stench must have been unbearable.

The Davy lamp, developed in 1815, improved safety, and the advent of electric light finally solved the lighting problem.

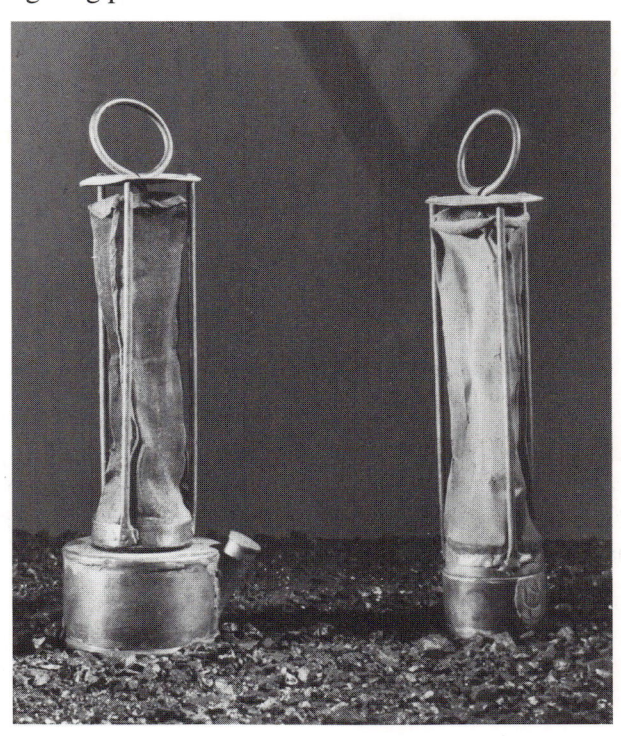

Two original Davy safety lamps.

Recognizing geological problems

Geological problems fall into two categories — gradual changes and sudden changes. Where a change is gradual, such as a seam thinning or splitting, data from boreholes in advance of the workings supplemented by measurements taken on working faces and development roads can be used to identify those areas where a seam becomes too thin or split to be worked.

Where the changes are sudden, as with washouts or faults, other methods have to be employed. For example, detailed studies of the strata exposed in roadways and in borehole sections might allow the paths of distributary channels to be plotted. Forecasting of fault positions is more difficult, and small faults with a displacement of less than 5 m are a major problem in detailed mine planning.

On an advance face the information about the geological conditions that lie ahead will be limited to that which can be deduced from any nearby workings or boreholes. So an advance face is a high risk layout, to be used in areas where uniform conditions are anticipated. On a retreat face the access roads expose any geological problems in the block of coal to be extracted, and the face can be scheduled to work back towards the main roadways without risks of interruptions. Consequently, the average daily output from retreat faces is some 25% higher than that from advance faces.

Block 4

Question 16

Why do geological disturbances result in so many problems for underground mining, yet have very little effect upon opencast mining?

For sound geological reasons, opencasting will always be carried out in preference to deep mining so long as the stripping ratio is favourable. Opencasting can work over 95% of the available coal and can exclude all the roof and floor sediments during production. Underground mining rarely recovers more than 70% of the available thick seams if it is to exclude roof and floor rock, and in addition small seams within the mined sequence cannot be effectively extracted.

The second part of Video Band 9: Energy Resources — Coal (from about 7¾ to 17 minutes) gives an overview of coal exploration and mining techniques. View this part now, and to check your grasp of it answer Question 17.

Question 17

(a) What geological information can be obtained from a single borehole through several coal seams?

(b) What extra geological information can be obtained from several more boreholes through the same coal seams?

(c) Which geological features relating to a single seam would a mining engineer wish to investigate prior to mining?

3.4 A modern coalfield case study: the Selby complex in North Yorkshire

3.4.1 Some general considerations

There is always a need to evaluate coal reserves in any particular area before a decision can be taken to work them. The major factor is the amount of workable coal present, but other factors such as purity and energy content also come into the calculation. The basic principle of tonnage calculations is straightforward. The tonnage is given by volume × density, and the volume of coal is controlled by seam area and seam thickness. Hence:

tonnage = seam area × seam thickness × coal density

However, many other factors come into play, as Figure 33 shows. *Seam area* is not the same as surface land area; the *dip* of the coal seam controls the land area required for mining. Any areas of former coal workings may need to have overburden removed as part of a larger-scale operation, but of course there is little or no coal left to work in these areas.

The economically viable area of a seam can often be reduced by faulting. In Figure 33, a large block of ground to the right of the shown fault is uneconomic for surface mining since the stripping ratio to the upper seam in that part is too great. In this case, detailed information from drilling has resulted in a substantial reduction of reserve.

Seam thickness can be proved by drilling, but even so, local geological factors may go undetected. In Figure 33, the seam split at the left side of the

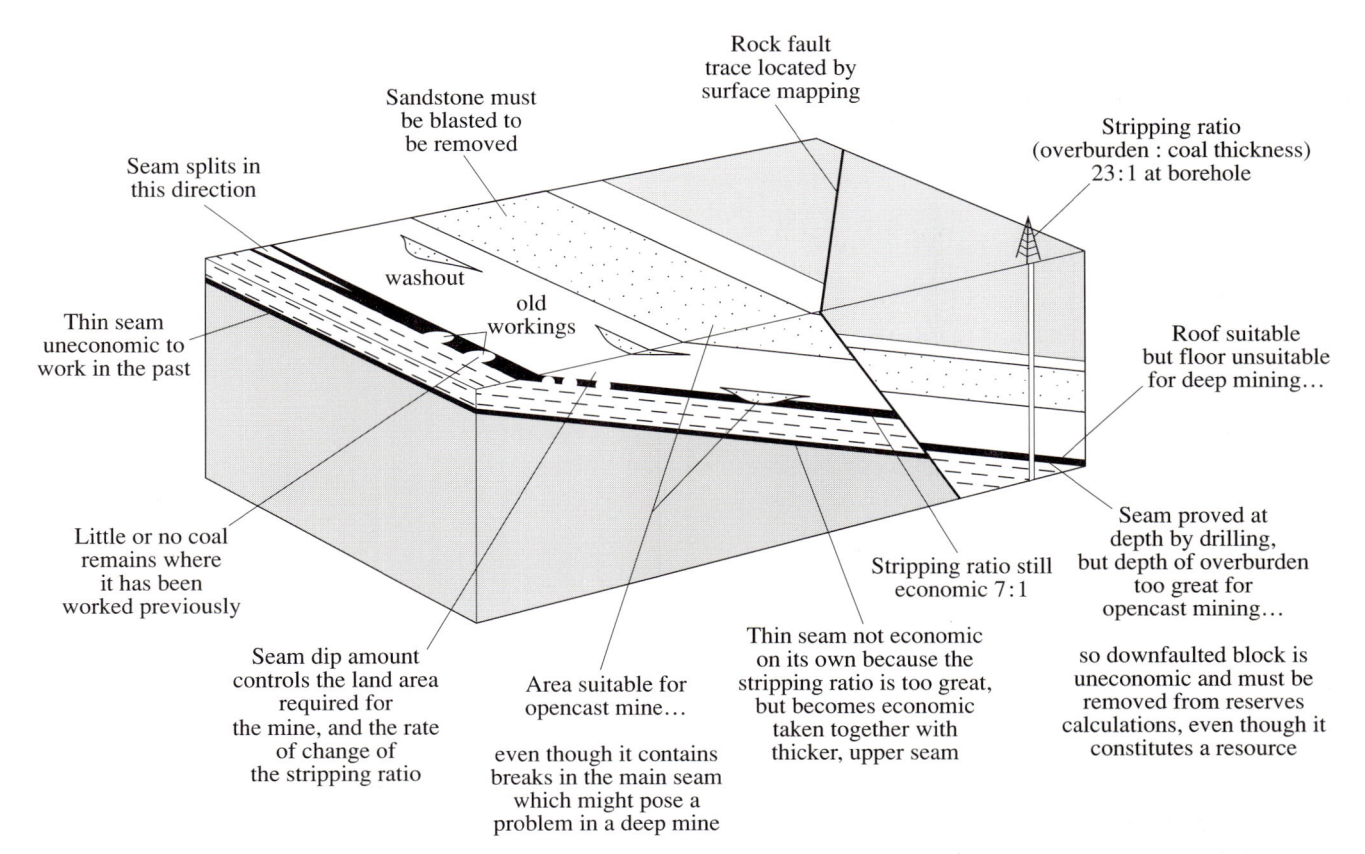

3 Finding and extracting coal

Figure 33 Some of the geological factors that influence reserve assessment and mine planning.

outcrop needs investigating. The nature of the rocks in the roof and floor may hamper deep mining within certain blocks. Where coal is to be deep mined, buildings and other surface structures, or special surface conditions, may mean leaving coal in the ground to support them.

The Selby mine complex in the Yorkshire coalfield, developed between 1976 and 1991, provides a case study of how the exploration and evaluation techniques can be applied as part of a modern coalfield evaluation. The principles outlined here for deep mining at Selby are in use for coal mining prospects all over the world.

3.4.2 The potential prospect at Selby

The development of the Yorkshire coalfield began in the 1840s with the exploitation of shallow seams along the western and northern edges of the exposed coalfield. As demand for coal increased, the working area was extended eastwards into the concealed section of the coalfield, underneath a cover of younger Permian limestones and sandstones. Progressively the mines became larger as the shafts were sunk to deeper levels and exploration became more expensive and more difficult.

Eastward development of the Yorkshire coalfield stopped for many years because it was believed that the principal coal seam (the Barnsley seam) deteriorated and split north-eastwards (Figure 34).

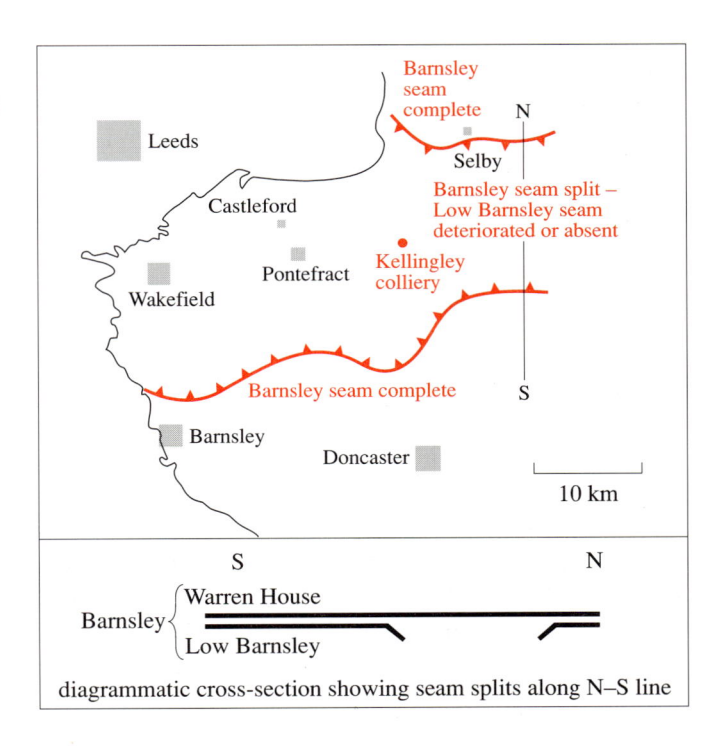

Figure 34 Diagrammatic map and cross-section of seam splitting in the Barnsley seam in the Selby region of the Yorkshire coalfield.

Exploration for the Kellingley colliery (11 km south-west of Selby town) in the 1950s led geologists to reconsider the evidence about the Selby area further to the north. They studied the available information — geological maps, old borehole records, and so on. Three old boreholes, drilled at the turn of the century (Figure 35), did not record any workable seams. The geological teams considered that these data were probably unreliable owing to potential loss of coal from the cores. If the conclusions drawn from the old boreholes were incorrect then the thick Barnsley seam, known to the west and south, might be complete to the north-east.

Re-drilling in the 1960s showed that it was in fact the early borehole data that were of poor quality, not the Barnsley seam. The split in the Barnsley seam did not continue north of Selby and there was a new and exciting prospect of workable coal over an area of 260 km² under the Vale of York north of the town of Selby (Figure 35).

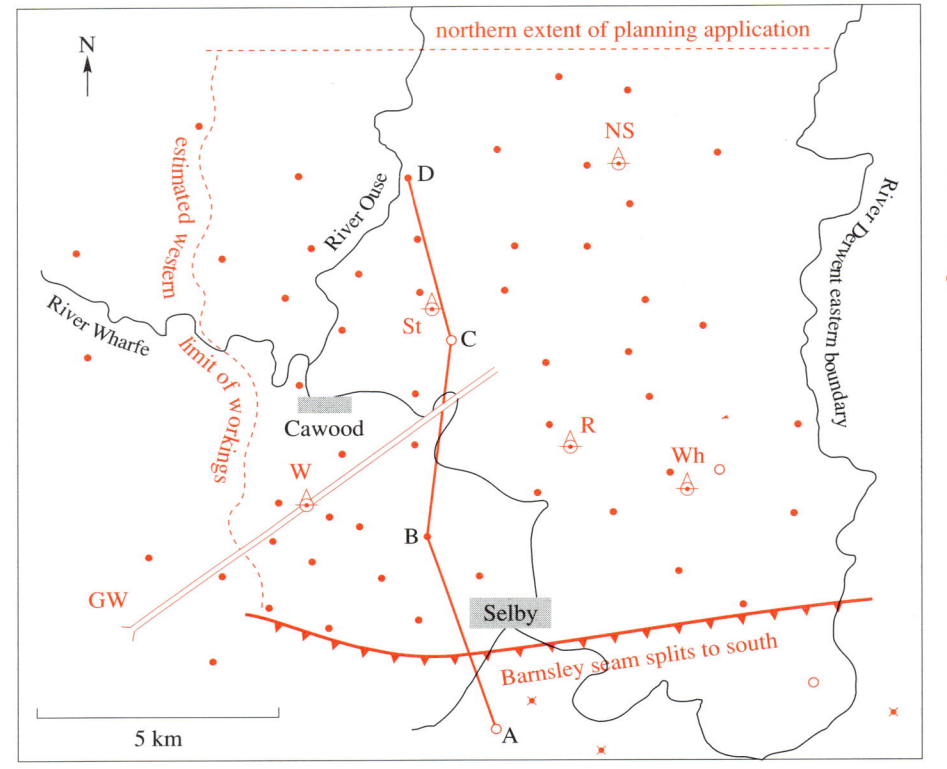

(a)

(b)

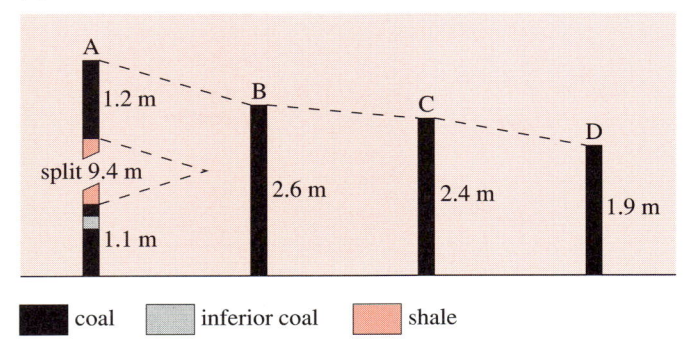

Figure 35 Exploration of the Selby coalfield. (a) The locations of four of the initial boreholes and many of the subsequent exploration boreholes. The planning limits of the Selby project are shown, together with the locations of the Gascoigne Wood mine entrance and the locations of the Wistow, Stillingfleet, North Selby, Riccall and Whitemoor mines. (b) A cross-section through the line of borehole A–D across the Selby district, showing the split in the Barnsley seam.

3.4.3 Exploration stage

Was the prospect sufficiently attractive to justify an exploratory drilling programme? During 1964–67 five boreholes were drilled around Selby; Figure 35 shows the locations of four of these. Improved coring methods located several thick seams and proved the existence of the very thick Barnsley seam, 1.9 m thick at its thinnest and increasing to 3.25 m near Selby.

Detailed assessment showed that there was 600×10^6 tonnes of reserves in the Barnsley seam. The coal was found to be very clean, with an ash content of 4–12%.

A major drilling programme was commenced in 1972, and eventually 68 deep boreholes were sunk to cover the whole field on a rough grid at intervals of 1.5–2.5 km (Figure 35). This drilling density is typical for the evaluation stages of a deep mine, but if the coal had been proved at shallower depths capable of economic opencasting, a much more extensive drilling programme at gridded intervals of only one or two hundred metres would have been planned.

3.4.4 Mine design and planning stage

Deep mining would only become economic if the annual output could reach or exceed 10^7 tonnes. Sufficient coal had to be proved with acceptably low risks of geological problems to ensure that the first 10 to 15 years of production would be trouble free. The drilling programme proved ten seams of workable quality lying at depths between 300 m and 1100 m. The Barnsley seam held at least 600×10^6 t; total coal resources measured in all seams amounted to 2×10^9 t.

The next step was to assess the most appropriate method of mining the coal.

○ The Barnsley seam varies in thickness from 1.9 m in the north of the area to 3.25 m in the south. From Figures 36 and 37, what would be the approximate stripping ratios at the Wistow and North Selby mines? Are these stripping ratios appropriate for opencast mining?

● At Wistow the seam is approximately 2.6 m thick (Figure 35b) and some 390 m underground (Figure 36). The stripping ratio here is 390 : 2.6 = 150 : 1. At North Selby it is greater, 1044 : 1.9 = 550 : 1. Both values are far greater than that mentioned in Section 3.3.1 as being the economic stripping ratio for opencasting.

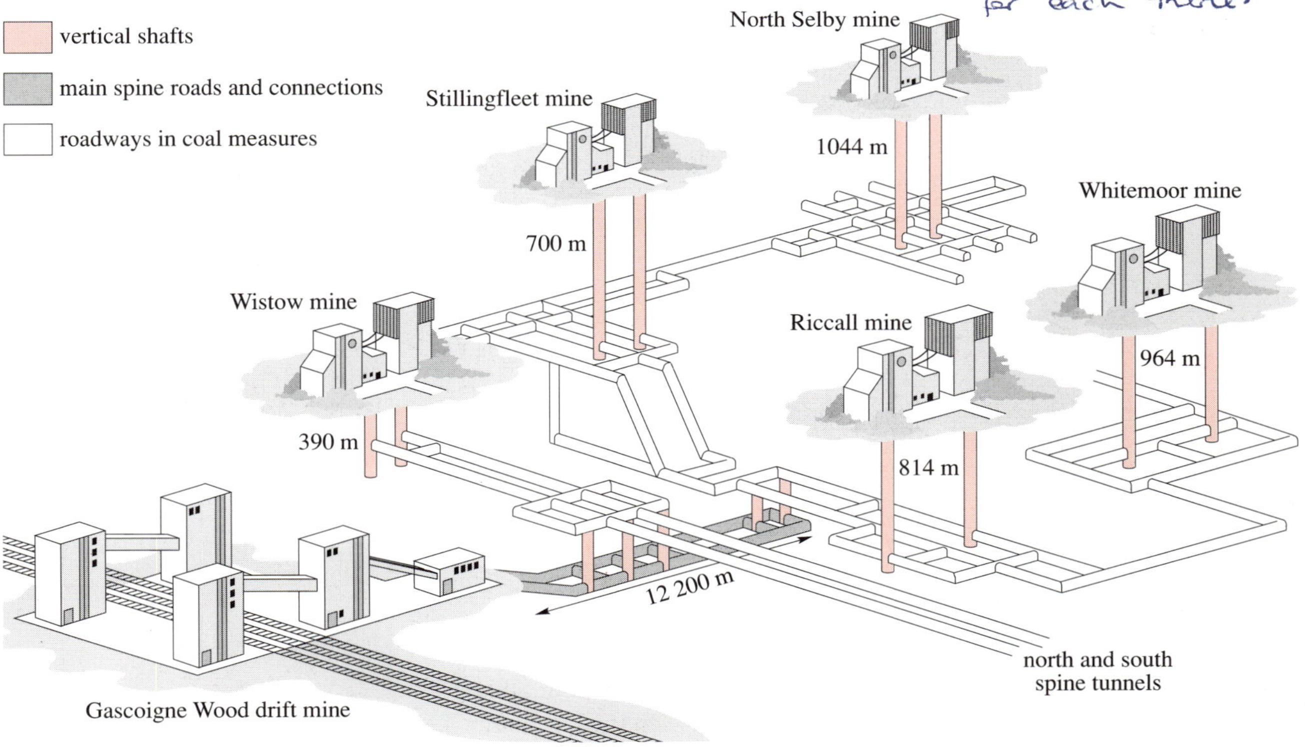

Figure 36 The Selby coalfield in 1992, showing the mine shaft, roadway and tunnel layout. *The depth from the surface to the Barnsley seam is shown for each mine.*

> **Dealing with water during tunnelling**
>
> Since Selby is a concealed coalfield, the shafts have to penetrate the rock sequence above the coal measures, which in Yorkshire includes Permian limestones and sandstones. There was also a potential problem with surface water in glacial deposits in the Vale of York.
>
> The latter problem was dealt with by starting the drift tunnels in an open cut, dewatered by 60 pumps. The Permian strata presented a more difficult technical problem, since they did not outcrop on the surface and were aquifers that could potentially yield over 10 000 gallons per minute.
>
> When shafts were started, a freezing plant was set up on the surface. Working like a domestic refrigerator, it pumped chilled brine down a collar of vertical boreholes around the margins of the shaft site. The brine froze the groundwater in the permeable formations into a solid plug, some 10 m in diameter. The shafts were then sunk through the ice and fitted with a waterproof lining before the ice was allowed to thaw.

The depth of the seams gave unrealistic stripping ratios, so the coal had to be extracted by underground mining methods. The scheme finally adopted was based on a long tunnel connecting working areas in different parts of the coalfield to a mine entrance at Gascoigne Wood on the western boundary. Shafts were sunk to serve each working area, providing ventilation and allowing access for workers and materials (Figure 35). These are now the five other mines named in Figure 35a.

A vital issue in the evaluation of the project was the problem of ground subsidence following extraction of the coal. The reserves lie under the flat Vale of York where the land is only some 7 m above mean sea level. This means that during high tides, water from the land drains must be pumped up into the tidal parts of the main rivers. Extensive ground subsidence would disrupt the land drainage system and could result in serious flooding of the rich agricultural land.

The proposed mining system therefore had to be designed to minimize these risks. Despite valuable reserves in ten seams, planning permission restricted British Coal to working only the Barnsley seam and to limiting the surface subsidence to a maximum of 0.99 m, with complete protection for the built-up areas of Selby and Cawood (Figure 35). These severe restrictions required much of the coal to be left in place in order to minimize surface subsidence, so that only 224×10^6 tonnes can be extracted.

- What proportion of the Barnsley seam can be recovered? Can the estimated *extractable* reserves in the Barnsley seam provide for the required 10–15 years of planned production?

- Recoverable proportions of the Barnsley seam are 224×10^6 tonnes/600×10^6 tonnes = 37%. At 10^7 tonnes extracted each year, the seam has a life of 22.4 years, well in excess of the planned production life of the area.

3.4.5 Mining in the Selby complex

In 1976, twelve years after the initial boreholes proved the first workable coal near Selby, planning consent was given to develop a £400 million colliery complex which started production in 1983 and reached its planned output in 1991. Note the very long lead times (over twenty years) needed to undertake both the exploration for a new venture and the investment and engineering works needed to achieve its completion.

The first face to come into operation was at Wistow in June 1983. It was 138 m long and then cost £4 million to equip. Since January 1991, coal has been produced at all of the five separate mines within the complex. Coal from ten retreat faces is transported over 12 km along a network of conveyor belts to a single coal-handling point at Gascoigne Wood. There the coal is washed to remove shale and mud, and transported by rail to four 'local' power stations at Ferrybridge, Eggborough and Drax A and B. Each of the four stations consumes 4–5 million tonnes of coal per year and each can generate electricity at a rate of about 2000 MW. The waste is also transported by rail to landfill sites in West Yorkshire.

In the mid-1990s, all five satellite mines within the Selby complex remained in full production, despite a major programme of closure for Britain's deep mines generally. The Selby complex produces about 48 600 tonnes of coal each day, or over 10 million tonnes of coal annually. This represents about 15% of all coal that was used in total for electricity generation in the UK in 1993. The remaining coal needed to fuel the Drax complex comes from other pits nearby.

Blocks of unworked coal have been left beneath the 900-year-old Selby Abbey, and as mining proceeds, these will be extended to protect a large area of Selby town centre and the bend in the River Ouse in the town.

3.5 Environmental aspects of coal mining

The two types of modern mining — deep mining and opencast — present significantly different environmental challenges. The environmental problems that surround opencasting are those of any large quarrying operation: sterilization of land and restoration of quarry sites, and dust and noise while operating. Mining waste is not a problem since it is used to fill the hole created. Deep mines can be less noisy and dusty, but cause subsidence and generate more permanent waste tips. Both types of mining share the fact that most of the coal they produce is used to fuel the generation of electricity — and burning either deep-mined and opencast coal causes atmospheric pollution. Section 5 is devoted to the effects of burning fossil fuels, so we shall consider here only the environmental impact of the mining operations themselves.

3.5.1 Opencast mining

Many environmental issues arise when opencast mining is considered, and proposals for new mines regularly arouse considerable local opposition. By its very nature, opencast mining has a major impact on the landscape, involving the digging of enormous pits with accompanying noise, dust and traffic movements, and destruction of mature landscape features. Increasingly in recent years the environmentally conscious public has used legal rights in the planning processes to oppose and sometimes prevent mining on sites where the environmental impact would be severe.

Many steps can be taken to minimize the nuisance impact of opencast sites. Topsoil is commonly formed into graded embankments around the boundaries of opencast sites as a baffle against visual intrusion, noise and dust. On-site vehicles can be fitted with effective silencers. To prevent dust being raised on-site, water bowsers spray haulage roads. Lorries leaving the site pass through wheel washers and typically their loads are sheeted over. Furthermore, individual opencast sites have lives limited to 5–10 years and engineers are required to restore former opencast sites to productive farmland, forestry or recreational use.

3.5.2 Deep mining

Deep mining operations have three significant areas of environmental impact — spoil heaps, subsidence and water pollution. Spoil heaps have always been the principal surface feature of underground mining operations but with the development of large mechanized collieries in the latter half of the twentieth century they have become increasingly intrusive on the landscape. Throughout the country considerable effort has been devoted in recent years to improving the visual appearance of mine sites and spoil heaps at existing collieries, by landscaping and by planting trees.

Mining subsidence

Subsidence is a potential hazard wherever deep mining is carried out. The major factors affecting the extent of subsidence are the *thickness* and the *depth of burial* of the seam extracted. The thicker the seam the greater the void left after mining, with consequently greater subsidence at the surface. The deeper the seam the more the effects of subsidence will be reduced through successive strata, so that the effects at the surface will be less although they will extend over a wider area.

How much land is affected by subsidence?

The area on the surface affected by subsidence is called the **area of draw**. This area is substantially larger than the area of worked coal underground (Figure 37) and is controlled by the **angle of draw** which is highly dependent on rock type and nature.

The amount of subsidence can be calculated roughly by using the formula:

$$s = \frac{4t}{\sqrt{d} + 4}$$

where s = depth of surface subsidence (metres),
t = thickness of the worked seam (metres),
d = depth to worked seam (metres)

Other factors affecting subsidence amounts include the amount of dip of the seam, the way the waste is packed behind the working face and the nature (particularly the joint pattern) of the overlying strata.

Subsidence starts within 24 hours of coal extraction, but the full effects are transmitted rather slowly upwards. It may be over 10 years before the surface is completely stable. The surface subsides in different ways near the edge of the area of draw (Figure 37). In the outer part of the area, the subsiding land surface is convex upwards, so the surface and any buildings on it are bowed outwards and stretched. Tension cracks will be common. As mining and extraction proceeds and subsidence occurs, the ground surface will become compressed. Different types of cracks and failures will develop and some cracks opened by tension may now be closed.

Any line of weakness in the surface rocks, such as an old fault, will tend to fail, giving rise to a stepped subsidence effect at the surface.

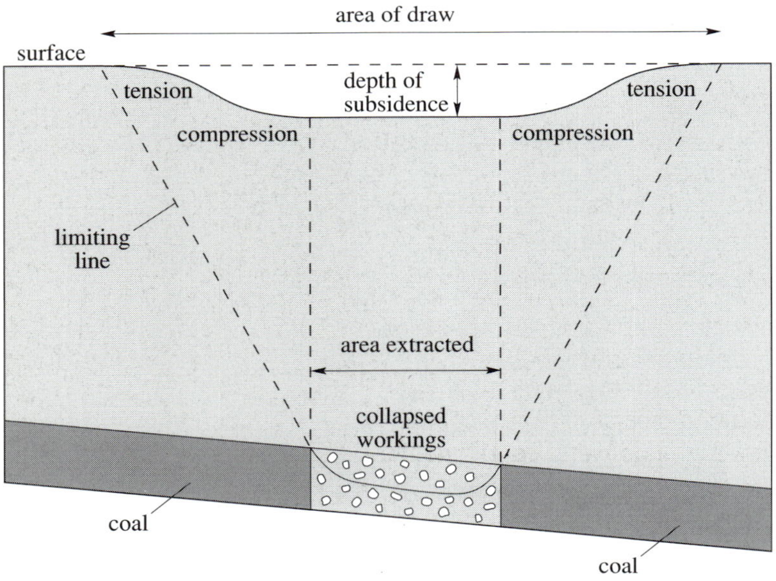

Figure 37 Extracting coal at depth produces subsidence at the surface. Within the area of draw, subsidence takes place in stages. Some parts of the surface are affected by tension and others by compression.

Activity 4

(a) Plot a graph to investigate the subsidence produced at depth by mining a seam of 3 m thickness under the Selby area.

Hint: You will need a calculator and a sheet of graph paper to do this. You will have to work out, using the subsidence equation in the box, the value of the amount of subsidence s for a seam thickness of 3 m for a range of working depths. We suggest that you use a range from 50 m to 750 m, at 100 m intervals, then record your calculated values of s and d in a table. Then use the graph paper to plot s on the vertical or y-axis against d on the horizontal or x-axis.

(b) From your graph, at what working depths can a 3-m seam be worked to produce less than the 0.99 m of surface subsidence demanded at Selby?

Certain vulnerable structures, such as dams, viaducts, and historical buildings, are protected by unworked pillars of coal left beneath them, but such protection may be extremely costly where it significantly affects the layout of the mine. The case study of the Selby mine highlighted a particular problem of subsidence affecting land levels in a low-lying area. At Selby, restrictions were imposed by government relating both to the visual impact of the surface developments at the mine sites and to minimizing the effects of ground subsidence in areas particularly liable to flooding or subsidence damage.

Water pollution from deep mines

The reasons for industrial water pollution and the nature of the pollutants were covered extensively in Block 3, Section 5. Both opencast and deep mines can potentially cause groundwater pollution since they usually lie well below the water table.

The main causes of minewater pollution are metal sulphates produced by the action of bacteria on pyrite (FeS_2) within the coal sequence. Before mining, groundwater flow through coal sequences was usually sluggish, groundwater was saline and generally slightly reducing chemically. During mining, most of the original salty water was extracted by pumping what was needed to keep the mine dry. Exposure of the coal-bearing rocks to moist air lead in turn to oxidation of pyrite within the sequence, helped by the catalytic action of bacteria. When mines close and pumping stops, the water table can rise again. Soluble pyrite oxidation products will be flushed into solution, generating waters polluted by iron, manganese and other metal sulphates. These solutions are extremely acidic and highly damaging to most forms of aquatic life.

Opencast mines represent only a transitory problem. Water is pumped out of them during working, and settling ponds are installed to ensure that only clean water leaves the site. The mined area is backfilled after use, and the water table generally equilibrates to that within the surrounding area when the former site is landscaped.

However, as deep mines close and pumping within them is stopped, they fill up with polluted water. The roadways and faces of abandoned coal mines form a very effective artificial network of underground water courses which can channel polluted water. Where this water emerges, surface streams and rivers can themselves become seriously polluted.

Preventive measures usually centre round preventing oxidation by excluding oxygen, water or bacteria from underground exposures. In old mines, pyrite oxidation products have already accumulated over centuries of exposure, and sealing old mineworkings to reduce water flow is impractical, requiring the water catchment area of the mine to be paved over.

A solution to this problem is not easy to find. It is expensive to continue pumping after a pit has closed down, and pumping would have to be maintained indefinitely to be effective. Routeing polluted minewaters into special treatment areas is impractical because, as we saw in Block 3, the route of groundwater flow is almost impossible to predict in anything other than general terms.

3.6 Coal reserves worldwide

Not all countries in the world have coal reserves, and there are several geological reasons for this:

1. *No sedimentary rocks*   Some countries, like Sweden, have almost no sedimentary rocks. These countries are naturally coal free, but such circumstances are relatively rare. All countries with large land areas, like USA, Russia, China and Australia, have sedimentary sequences, often in thick sedimentary basins, and these countries have substantial coal reserves.

2. *No post-Devonian sediments ever deposited*   Countries where no post-Devonian rocks were ever deposited have no coal whatsoever. Norway has a thick sequence of sediments of Lower Palaeozoic age in a basin around Oslo, but this sequence does not contain coal because these rocks are simply too old.

3. *Coal-bearing rocks eroded*   Countries may have no coal-bearing rocks because although they were once deposited they have now been eroded away. Ireland once contained a substantial sequence of Upper Carboniferous coals, almost all of which has now been eroded. The Republic of Ireland is left with about one-twenty-thousandth of the total coal reserves in these islands, but in the distant geological past Ireland may have contained almost as much coal as Britain.

4. *Coal-bearing rocks too deep*   Much of the world's coal comes from rocks of late Carboniferous or Permian age. Some countries have coal-bearing Carboniferous or Permian rocks that are buried so deeply under younger sedimentary sequences that they cannot be worked economically. The Netherlands is a good example.

5. *Sediments of right age but wrong type*   Some countries have rocks of Carboniferous or Permian age but of the wrong type of sediments. This might be because the land area lay at the wrong palaeolatitude in Carboniferous or Permian times (coal is a product of warm to hot climates), or because these rocks were deposited in the deep sea, rather than on coastal plains and deltas. The Carboniferous of South Africa and the Permian of Britain are particularly good examples, as the Permian of South Africa and the Carboniferous of Britain both contain substantial coal sequences.

Britain and Western Europe are fortunate in having extensive Carboniferous coalfields. Figure 38 is a **palaeogeographic map** of the present-day Atlantic region, i.e. a map of the region as it appeared in late Carboniferous times, about 300 Ma ago. Maps like this are compiled from many years of careful research into the Carboniferous geology of Europe. As you can see, there are some profound differences between the region's present-day geography and that of the past.

3 Finding and extracting coal

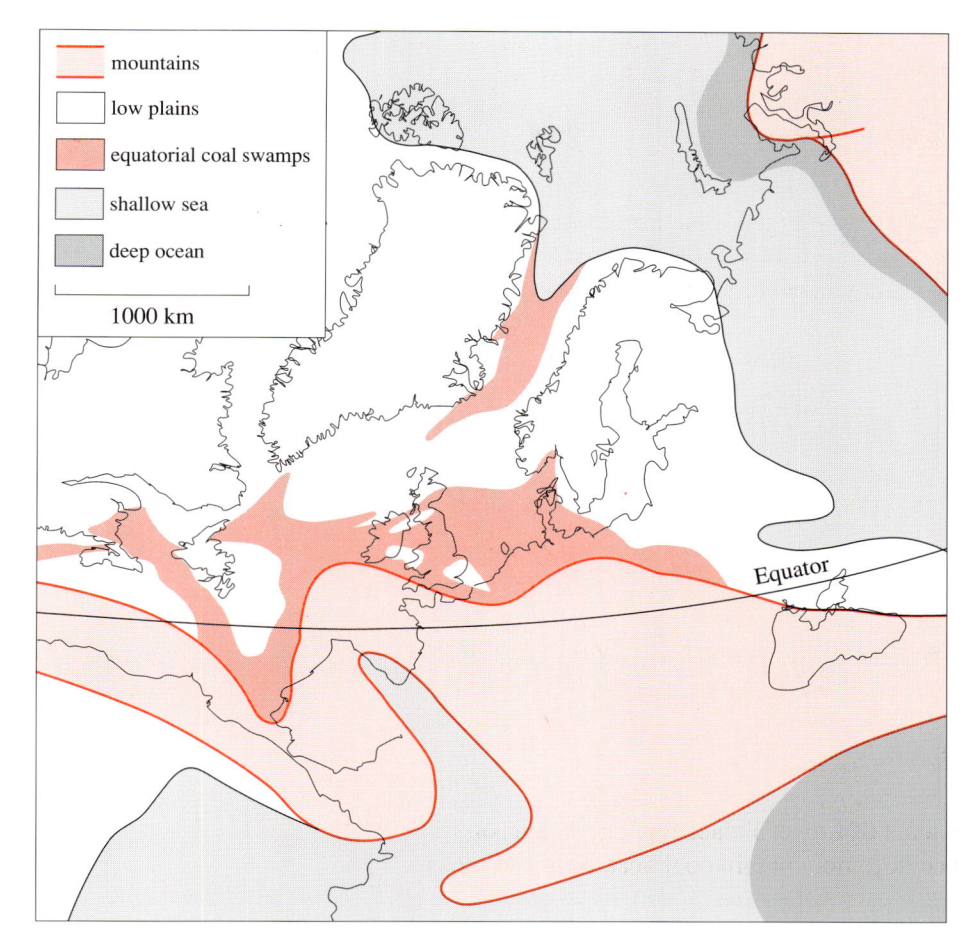

Figure 38 A palaeogeographic map of the present-day Atlantic region in late Carboniferous times, about 300 Ma ago. Most of central Europe was mountainous, bounded to the south-east by a major ocean. Coal swamps developed in an equatorial belt stretching from eastern USA through northern continental Europe to the site of the Black Sea.

- What does Figure 38 tell us about (a) the Atlantic Ocean, (b) the Equator in Carboniferous times?

- The Atlantic Ocean did not exist in Carboniferous times, and the Equator lay through western and central Europe during late Carboniferous times.

Carboniferous Britain and north Europe formed an equatorial plain sandwiched between an area of mountains developing to the south and an extensive area of land of modest relief to the north. The nearest sea lay away to the east in Russia, and to the south in northern Africa. Tropical swamps developed across Britain and Ireland, through the southern North Sea into Belgium, the Netherlands and northern Germany and east into Poland. The same conditions extended westwards into eastern USA, and also towards a major arm of the sea between Greenland and Scandinavia. Coal formed across the whole of this belt.

However, if you compare Figure 38 with Figure 39 you will see that Carboniferous Coal Measures are no longer preserved over *all* the area once covered by coal swamps. Figure 39 is a map showing where Carboniferous rocks and coal measures either outcrop at the surface or exist at depth, buried under younger rocks. Normally in sedimentary basins, older rocks are buried under younger ones, as is the case in the Netherlands, the North Sea and east Yorkshire. Where there has been uplift due to tectonic activity, younger rocks may be eroded away, exposing Carboniferous Coal Measures . This has happened in the Ruhr area of Germany, in Poland, in parts of Lancashire and Yorkshire and in the Scottish Midland Valley. More extreme uplift leads to the erosion of the coal-bearing sequence, too, as has happened in the Pennine area and Ireland.

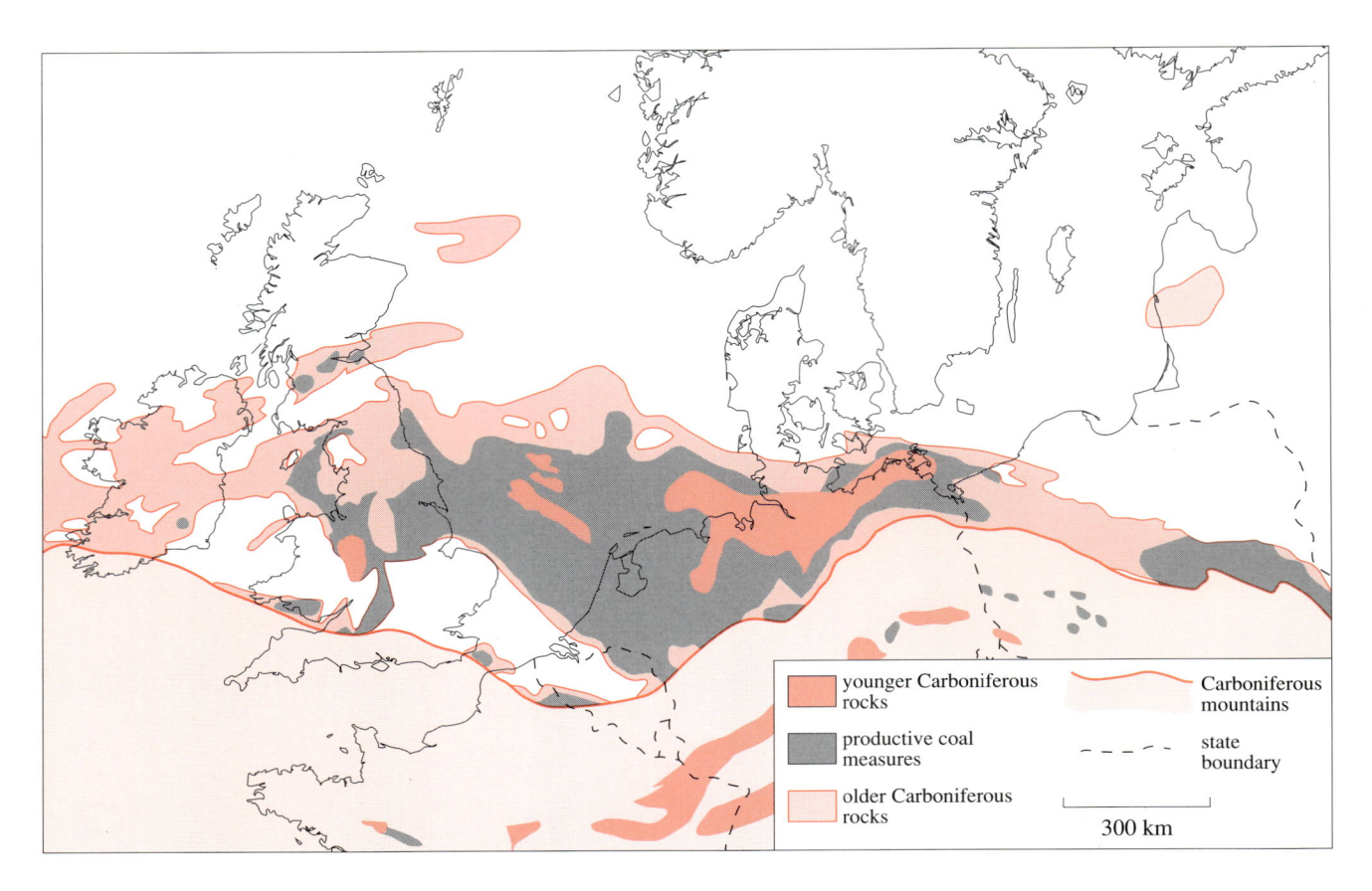

Figure 39 A map of present-day north-west Europe, showing where Carboniferous rocks are preserved, either at the surface or covered by younger rocks.

Map legend:
- younger Carboniferous rocks
- productive coal measures
- older Carboniferous rocks
- Carboniferous mountains
- state boundary
- 300 km

⬤ Use the *Postcard Geological Map* and Figures 39 and 40 together to answer these three questions:
  (a) Which areas of Britain have Carboniferous rocks at the surface (ignoring glacial deposits)?
  (b) Which parts of Britain could not contain coal because the surface rocks are too old?
  (c) Which parts of Britain have extensive areas of post-Carboniferous cover and form potential concealed coalfields?

○ (a) Carboniferous rocks are located at the surface in the Central Valley of Scotland, north-east and north-west England, the industrial belts of Lancashire and Yorkshire, the English Midlands, South Wales and Devon.
  (b) Surface rocks are too old for coal to be present in all of Scotland north of the Central Valley, the Southern Uplands of Scotland, the Lake District, almost all of north and central Wales and Cornwall.
  (c) Concealed coalfields *could* occur in almost any part of England south-east of a line from Teesside to Exeter. In fact, because of the nature of the Carboniferous geography, Upper Carboniferous coal sequences were never deposited over a large area of East Anglia and south-eastern England that was upland in Carboniferous times (the 'London–Brabant massif').

Question 18

Is there a different geological explanation for the lack of Carboniferous coal in East Anglia and the lack of Carboniferous coal in the Pennines of northern England?

3 Finding and extracting coal

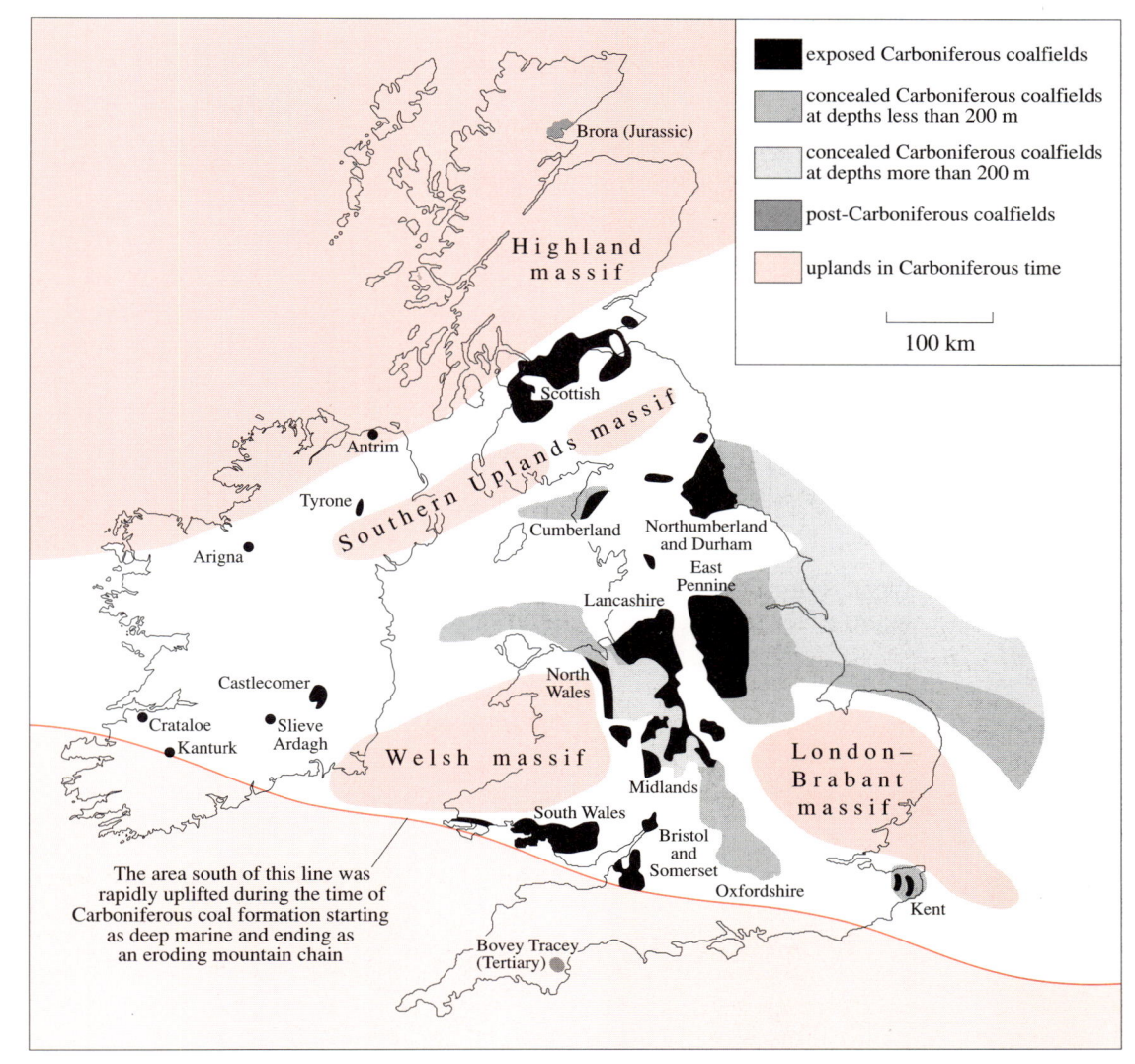

Figure 40 The location of the British and Irish coalfields and their offshore extensions. The shaded and named areas were uplands during late Carboniferous times, and hence no coal was deposited on them. Two small, presently uncommercial deposits of Jurassic and Tertiary coal are also shown at Brora and Bovey Tracey. Jurassic coal has also been worked in the past in north Yorkshire.

3.6.1 UK coal reserves

Coal mining in Britain began in the coastal coalfields of Scotland, north-east England and south Wales (Figure 40), but the reserves of these regions have become depleted and deep-mining activity there has almost stopped. Present-day deep mining is almost exclusively limited to the East Midlands and Yorkshire regions, although opencast sites still work coal in most of the coalfields.

Because of the intensity of late Carboniferous tectonic earth movements in southern Britain and the high heat flows associated with them, the South Wales, Somerset and Kent coalfields contain high-rank coking coals and anthracites. These are the most useful British coals from an energy-production standpoint yet at the same time are the most difficult to mine because of the deformed nature of the coal strata.

In the north of the country the coal is of lower rank. Erosion has removed many hundreds of metres of Carboniferous strata, so that the central and northern coalfields are now separated into isolated fields. The coastal coalfields of northern England and the Midland Valley of Scotland extend eastwards under the North Sea and coal has been worked there in the past, but the large resources off the Yorkshire–Lincolnshire coast have not yet been mined.

Coals younger than Carboniferous do occur in Britain, but they are not of any significant economic importance. A small Jurassic coalfield, part of a larger off-shore field, was formerly worked at Brora on the north-east coast of Scotland. In 1982 substantial Tertiary lignite deposits were discovered on the eastern shores of Lough Neagh in Northern Ireland.

In 1993 British Coal estimated that there was a *total stock* of 190×10^9 tonnes of coal of all types in place in the UK, in seams over 0.6 m thick and less than 1200 m deep. Much of this is too thin and too deep to be worked commercially using existing technology. In 1993, British coal *reserves* were estimated to be 3.3×10^9 tonnes of anthracite and bituminous coal, and 0.5×10^9 tonnes of sub-bituminous coal and lignite. This represents 0.4% of total world coal reserves of all types.

⬤ What percentage of Britain's total coal and lignite stock was classed as 'extractable profitably and legally under existing conditions' (i.e. reserves, as defined in Block 1) in 1993?

◯ Britain's total coal reserves of all types amounted to 3.8×10^9 tonnes in 1993. This means that 2% (3.8×10^9 divided by 190×10^9 tonnes) of total stock was classed as 'extractable profitably and legally under existing conditions'.

3.6.2 Europe's coal reserves

Figure 39 shows the extent of Carboniferous coal deposits in north-west Europe, and Table 5 shows coal production from the 12 countries that in 1993 were members of the European Community (now the EU), ranked in order of greatest tonnage. Britain produced the most hard coal, followed by Germany, then Spain and France. As you can see, Britain and Germany lie centrally across the north European coal belt. The Carboniferous coal produced in Spain and France is high-grade but comes from smaller basins than those found further north in Europe.

Table 5 Coal production from European Union countries in 1993

EU country	Coal production in 1993		
	All ranks ($\times 10^6$ t)	Hard coal (anthracite and bituminous coal) ($\times 10^6$ t)	Brown (sub-bituminous) coal and lignite ($\times 10^6$ t)
Germany	279.7	57.9	221.8
UK	68.3	68.3	none
Greece	55.9	none	55.9
Spain	32.1	14.4	17.7
France	10.7	9.3	1.4
Italy	0.9	0.2	0.7
Others (Portugal, Belgium, Ireland, Netherlands, Denmark, Luxembourg)	0.6	0.6	none

Source: BP.

Table 5 also shows that many EU countries produce brown coal in large quantities. In 1993, Germany, Greece, Spain and Italy produced more low-rank brown coal than higher rank coals, as indeed did Czechoslovakia and Hungary. This is not Carboniferous coal, but coal that formed during Tertiary times, about 20 Ma ago. As Figure 41 shows, the North Sea reached far into Germany in middle Tertiary times. Coastal swamps around the North Sea were densely forested and protected from the sea by a substantial sand bar. Peat accumulation kept pace with basin subsidence and prevented the sea from entering the coal basin.

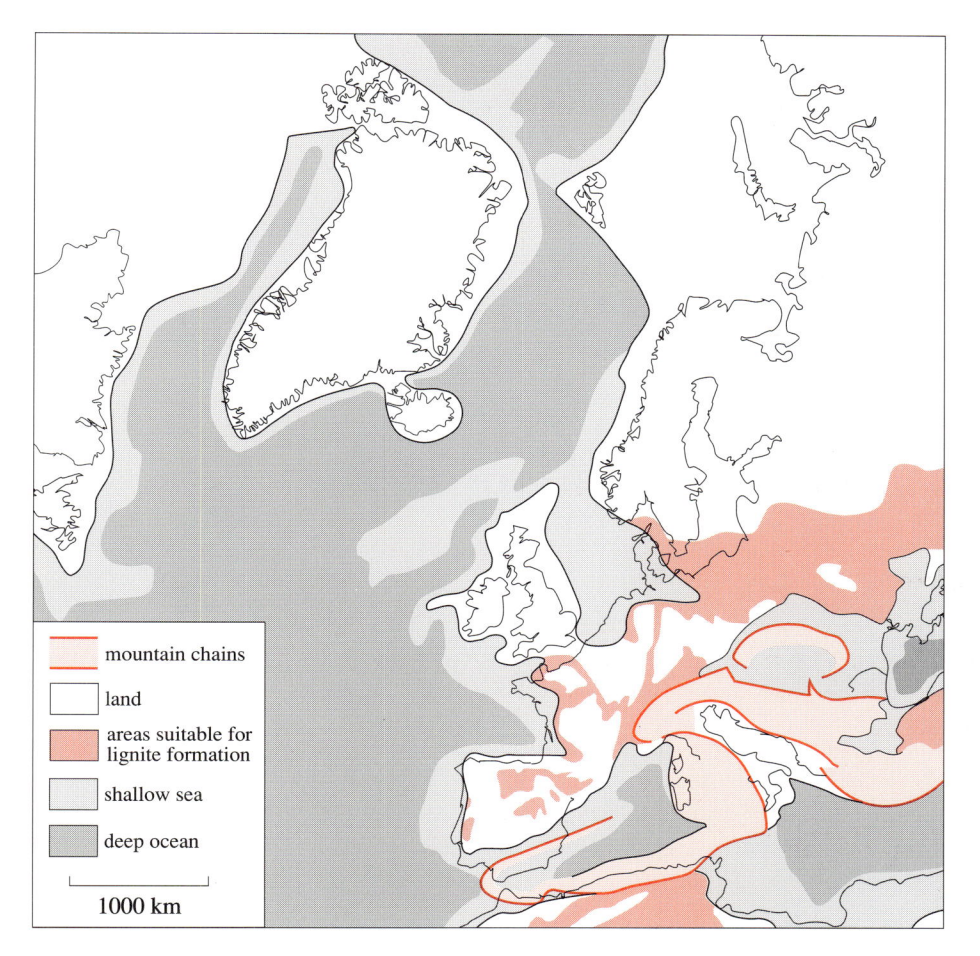

Figure 41 A palaeogeographic map of Europe and the Atlantic in Tertiary times, about 20 Ma ago. Symbols as for Figure 38. By mid-Tertiary times, the Atlantic Ocean had formed and the distribution of the continents was much as we see them today. Britain was still connected to mainland Europe by a land bridge, but the North Sea extended deep into Germany. The countries of Germany, Poland and the Ukraine, as well as areas of France, Spain and Austria, were suitable sites for dense forest growth, with subsequent peat and brown coal formation.

In 1993, Western European coal reserves were estimated to be 29.3×10^9 tonnes of anthracite and bituminous coal (mostly Carboniferous), and 67.6×10^9 tonnes of sub-bituminous coal and lignite (mostly Tertiary). This represents 9.3% of total world coal reserves of all types. Eastern European coal reserves (excluding Russia) were estimated in 1993 to be 32.2×10^9 tonnes of anthracite and bituminous coal, and 42.3×10^9 tonnes of sub-bituminous coal and lignite, representing 7.2% of total world coal reserves of all types. Poland alone has almost the same reserves of anthracite and bituminous coal as the whole of the EU countries taken together.

3.6.3 World coal reserves

Figure 42 shows the worldwide distribution of coal deposits. The known major areas are principally in the northern hemisphere. With the exception of Australia the southern continents are relatively deficient in coal deposits.

There are two broad belts of coalfields. A broad chain of large coalfields of Carboniferous age extends from the USA through Western and Eastern Europe, Russia and into China. A second chain of Permo-Triassic coalfields is found in the southern continents — South America, southern Africa, Australia and Antarctica.

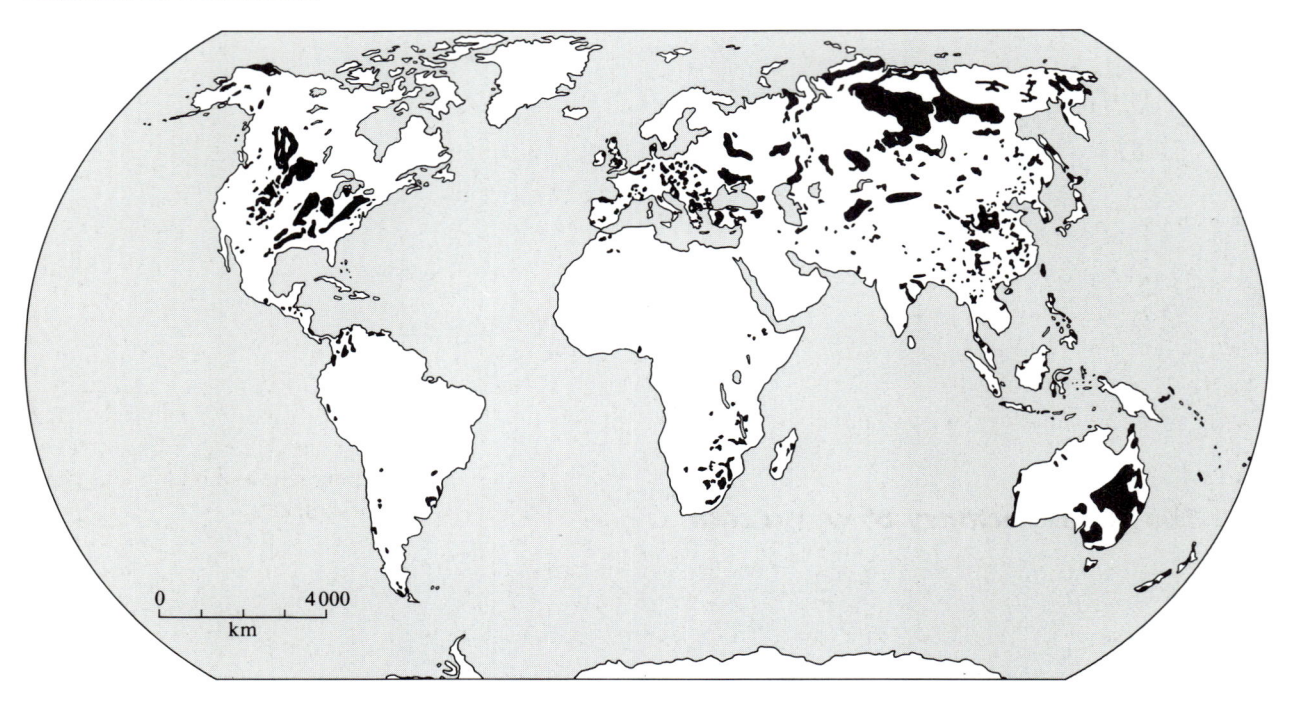

The coals of the southern continents are rather different from Carboniferous coals. They have a distinctive, temperate-latitude flora not found in the northern hemisphere. They span in age from the Permian to the Triassic periods. Mesozoic–Tertiary lignites are also important global sources of coal. The vast coalfields of western USA and Canada are Mesozoic in age.

Figure 42 A world map showing the distribution of known coal deposits of all ages.

Now watch Video Band 9: *Energy Resources — Coal* again. This will serve to review the main points of the Block. You should concentrate this time on the last part of the video, about coal mining in North America. Check your understanding of this part by answering Question 19.

Question 19

(a) Why is the coal in Sparwood, British Columbia, such a high-quality coal?

(b) Suggest three reasons why the site at Wabamun, Alberta, is suitable for opencast working.

In 1993, world proven coal reserves were estimated at 1039.2×10^9 tonnes, of which slightly over half (50.2%) was anthracite and bituminous coal and the rest (49.8%) was sub-bituminous coal and lignite.

Figure 43 shows the breakdown of world reserves by continents. North America has 24% of total world coal reserves, Asia and Australasia together have 30%, Eastern Europe 30%, and Western Europe 9%.

3 Finding and extracting coal

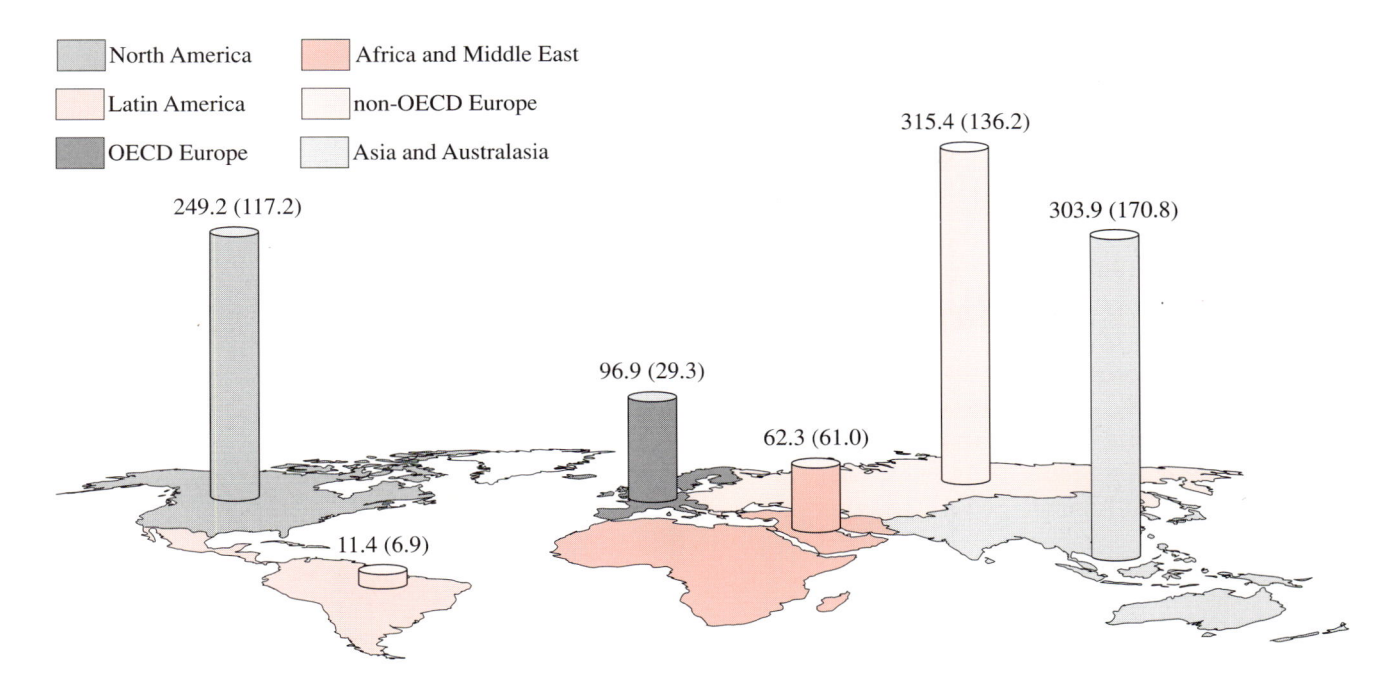

[Legend: North America; Latin America; OECD Europe; Africa and Middle East; non-OECD Europe; Asia and Australasia]

- 249.2 (117.2) — North America
- 11.4 (6.9) — Latin America
- 96.9 (29.3) — OECD Europe
- 62.3 (61.0) — Africa and Middle East
- 315.4 (136.2) — non-OECD Europe
- 303.9 (170.8) — Asia and Australasia

In Latin America, coal provides only 6% of energy, whereas in Western Europe it provides 20%. The greatest demand for energy from coal at present comes from Asia and Australasia, where coal supplies about 47% of energy needs.

Figure 43 World coal reserves proven by the end of 1993. Amounts are in thousand million tonnes, with the total of anthracite and bituminous coal in brackets.

3.6.4 The life expectancy of world coal

For any resource like coal, the ratio of *proved reserves* in the ground (R) to *the annual amount of production* (P) is a useful indicator of the 'state of health' of the reserves. The **R/P ratio** changes from month to month as new reserves are proven and as production rates rise and fall. An R/P ratio of 30–40 is normal for an extractive industry which has a long time-lapse from exploration to extraction. (In Block 2, a similar ratio for sand and gravel extraction was termed the *land bank* and was often as small as 5–10 years.) What then are current world coal R/P ratios?

- In 1993, 4.48×10^9 tonnes of coal were produced worldwide. What was the global R/P ratio for coal in 1993?

- Dividing this figure into the world total coal reserves quoted above gives an R/P ratio of 232.

However, the R/P ratio has dimensions of tonnes divided by tonnes per year, which calculates out as *years* ($t/t\,y^{-1} = 1/y^{-1} = y$). Therefore the R/P ratio also gives information about the lifetime of any fossil fuel, seen from today's perspective; the same as the *reserves lifetime* that we defined in Section 2 of Block 1.

- Why (recalling the discussion from Block 1, Section 2) can reserves lifetimes remain reasonably constant for decades?

- Reserves lifetimes reflect the pace of exploration and evaluation, versus extraction, that allows reserves to be maintained for the foreseeable future.

World coal could last comfortably into the twenty-third century without any further reserves being identified, *if production were to be stabilized at present-day levels.* Of course, stabilizing world production is neither possible nor desirable, so the R/P 'lifetime' must be interpreted with extreme caution. As we saw in Block 1, production of any resource is a complex interplay of location, economics and politics. British coal production in the early 1990s provides a good example.

3.7 Substitution and economics of coal production in Britain in the early 1990s

Coal has always been considered to be a product with *high place value*, but over recent years the advent of bulk transport ships and huge stripping and mining plant has changed coal economics drastically. Most coal is still used in the region where it is produced, but now coal can be produced in North and South America and Australia, where opencast mines work thick seams close to the surface, and transported to Britain for less cost than producing Britain's own coal from deep mines.

Figure 44 shows some UK coal production and consumption figures, drawn from British Coal's annual report for 1992–93. It shows tonnages of coal per year for a host of different situations. Try these questions to help you understand Figure 44:

- In what year did total coal production peak? How much coal did we produce in that year? In 1993, did we import more coal than we produced from opencast mines?

- Total coal production peaked in 1955, when we produced 225 million tonnes — 213 Mt deep-mined, 12 Mt opencast. In 1993, we imported 3 Mt of coal more than we produced from opencasting.

The information in this graph is of great economic significance. The *demand for coal in Britain has fallen steadily since 1955*, as the *total consumption* curve on Figure 44 shows.

- Why is this?

- Because despite the fact that increasing amounts coal were used for power generation between 1955 and 1980, the combined market for coke and domestic coal reduced by 80% over the same period. Since then demand for domestic and coking coal has remained steady, but demand for coal for power generation has declined sharply.

Coking coal was used extensively within the British steel industry; the decline in demand for coking coal in Britain between 1955 and 1980 reflects the decline in the British steel industry over that period. We now import rather than manufacture most of our steel, and British steel has been substituted by foreign steel. It is a good example of what was defined in Block 1, Section 2 as *cross-elasticity of demand*.

The demand for domestic coal reduced over the same period because of a change in home heating preferences. Domestic coal consumption was steady from 1945 to 1960, but fell sharply from 1960 to 1980. This corresponds to the period over which North Sea gas became widely available. Gas fires and boilers are considerably easier and less messy to run in a home than coal fires and solid fuel stoves. In this case, coal produced in Britain as a fuel was substituted by gas produced in Britain as a fuel. This is largely *substitution through technological advance* (Block 1, Section 1.4).

3 Finding and extracting coal

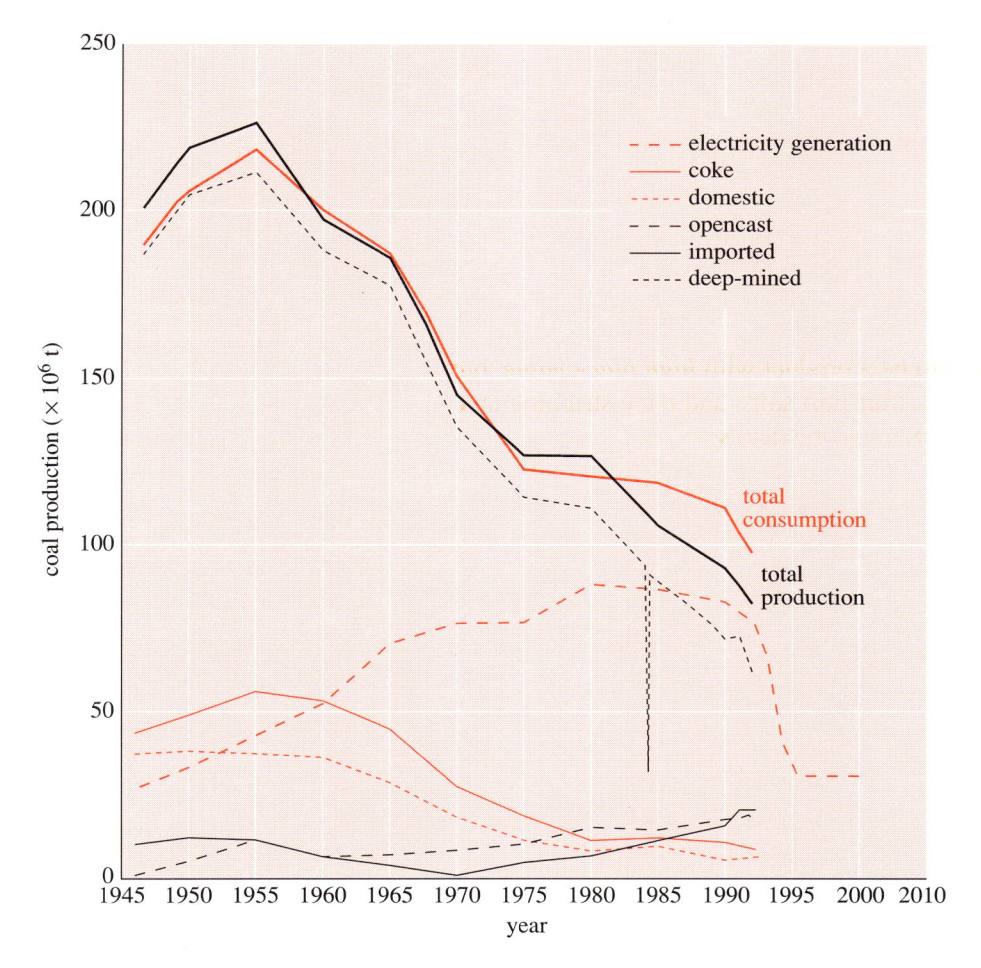

Figure 44 UK coal production and consumption figures for the period 1945 to 1994, and coal-use projections until the end of the twentieth century. (The miners' strike accounted for the drop in production in 1984.)

● From Figure 44, describe the changes in amount of coal used for electricity generation from 1950 to 1993.

○ Coal demand for electricity generation increased steadily until 1965, then somewhat erratically until 1980. Since 1980, demand for coal for electricity generation has fallen; steadily at first, then sharply since 1990.

By the late 1970s, a rationalization of the steel industry and a change away from coal as the main source of domestic heating had taken place. Yearly total consumption of coal in Britain had steadied at around 120 million tonnes. In the period from 1975 to 1985, increased production of coal from British opencast mines and increased imports of cheap opencast coal from abroad were offset by a reduction of coal supplied from Britain's deep mines. This is an example of *substitution for economic reasons* (Block 1, Section 1.4); opencast coal worked in Britain or transported from abroad was cheaper than British deep-mined coal of the same quality.

In the 1990s, British electricity generation industry was privatized and the two new companies, PowerGen and National Power, moved away from coal to natural gas as their preferred fuel. The initiative was dubbed at the time 'the dash for gas'. As Figure 45 shows, coal demand in England and Wales is expected to drop sharply in favour of gas over the 1990s. Coal used to generate electricity dropped from over 80 million tonnes in 1990 to some 30 million tonnes in 1994. The main reasons for this change can perhaps be classified as *substitution for reasons of political and economic preference*.

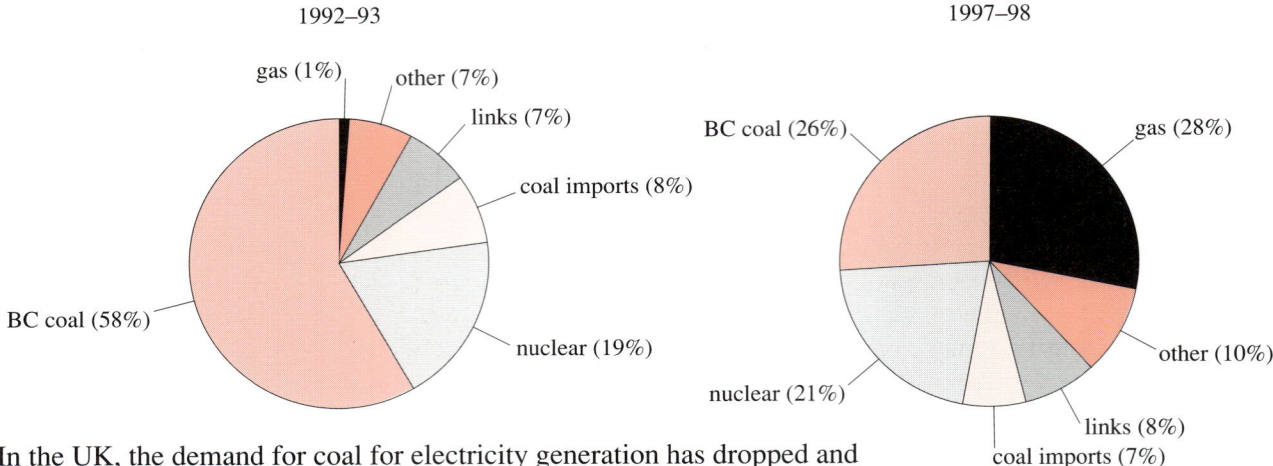

In the UK, the demand for coal for electricity generation has dropped and can now easily be supplied by opencast production at home and coal imports from abroad. As a direct consequence, the British deep-mined coal industry virtually closed down in the early 1990s. In 1984 there were 170 deep mines; by 1994 there were only seventeen. Despite a drop by almost one-half in the cost of deep-mined coal from 1985 to 1993, and a more than doubling of underground productivity over the same period, there will probably be less than twenty deep pits working coal in Britain throughout the 1990s.

Many of these are newer pits working thick seams, like the Selby complex and new developments at Asfordby in Leicestershire and Maltby in Yorkshire. These newer pits were commissioned in the 1980s, when the dramatic decline in demand for coal could not have been foreseen. Nevertheless, these new pits can produce coal more cheaply than older pits which have less advanced mining technology. Changing fortunes and long lead times in the coal industry have inevitably forced the closure of many older mines, bringing hardship once again to established mining communities.

The plight of UK coal illustrates dramatically how vulnerable the concept of reserves is to economic and political change. Resources are physical entities which exist for geological reasons. Whether they constitute reserves is a technological, economic and political question. The British coal reserve situation is summed up in Figure 46.

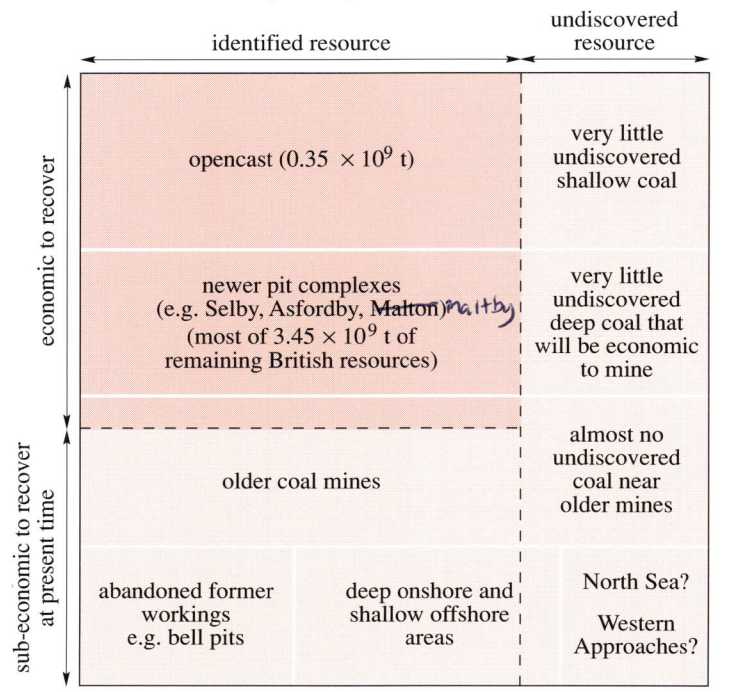

Figure 45 The expected balance of fuels supplying electricity in England and Wales over the 1990s. 'Links' represents imported electricity, mainly from Scotland and France; BC, British Coal.

Figure 46 UK coal resources in the mid-1990s. This diagram should be compared with Figures 29–31 in Block 1.

However, changes in the status of reserves and working coal areas has resource implications. Unworked areas of deep or potentially opencast coal can be preserved almost indefinitely, but closing down deep mines sterilizes coal, through flooding and through subsidence. Changing fuel means losing fuel, too.

From today's perspective the electricity generation sector worldwide cannot manage without coal. In the past, significant economic and technological developments provided the opportunities for coal use. In the future, the need to burn coal cleanly will control the amount of coal that will be used. Despite the demise of British deep-mined coal, the global future for coal looks more optimistic than for any of the fossil fuel alternatives, simply because there is more of it.

3.8 Summary of Section 3

1. Coalfields can be classified as either exposed or concealed, depending whether or not the coal-bearing rocks are hidden by younger strata. In most coalfields, mining commenced in the shallower exposed regions and has gradually extended into the deeper parts of the concealed coalfields.

2. Surface outcrops of rock can tell us much about whether there is a likelihood of coal at depth. Remote areas of the world are difficult to access on foot or by vehicle, so initial surveys often use data acquired from satellites or aeroplanes.

3. There are several effective and widely used exploration techniques. Drilling boreholes is expensive but is the only way to be absolutely certain of rock sequences, thicknesses and ages. Geophysical logging records the nature of the strata located in the boreholes without the need to take cores. The coals themselves are cored for chemical analysis, though.

4. Most deep coal is extracted nowadays using the longwall method and mechanized systems. Such coal-cutting systems are very inflexible and are incapable of negotiating many of the geological variations likely to be met. An understanding of the probable geological variations underground is therefore essential, both for selecting working areas and maintaining the continuity of face operations.

5. Surface or opencast mining is a flexible, cheap system that is currently producing most of the world's exported coal. Opencast mining is limited only by the stripping ratio of overburden to coal.

6. Opencast mining results in short-term local environmental disturbances; underground mining provides longer lasting problems, owing to spoil heaps, subsidence and drainage.

7. The oldest coalfields are of Carboniferous age which formed in tropical latitudes. The Permian or Triassic coalfields of the southern hemisphere originated in temperate latitudes. Brown coals make up half of the world's coal reserves; mainly they are Jurassic to Tertiary in age. Extensive Tertiary deposits of low-rank coal are worked in Eastern Europe.

8. Coal reserves can be calculated roughly on a straightforward basis of multiplying the area of occurrence by thickness of seam and the density of the coal. However, many geological factors affect this calculation. For accurate reserve estimations, detailed geological and geophysical surveys and drilling programmes must be undertaken. In 1993, the world had reserves of 1039 billion tonnes of coal, which at 1993 rates of production would last beyond the year 2229.

9   Demand for coal has fallen in Britain over the last twenty years, because of technological, economic and political factors. Britain's current coal demands could be met without using any deep-mined coal.

Question 20

Figure 47 illustrates a simplified geophysical log. It consists of the natural gamma ray and density logs from a coal-bearing sequence in which the following rock types occur: coal, shale, sandstone, marine band and seaearth. Which of the nine bands, A–I in this sequence, are coals? (The rock type boundaries are marked in the column.)

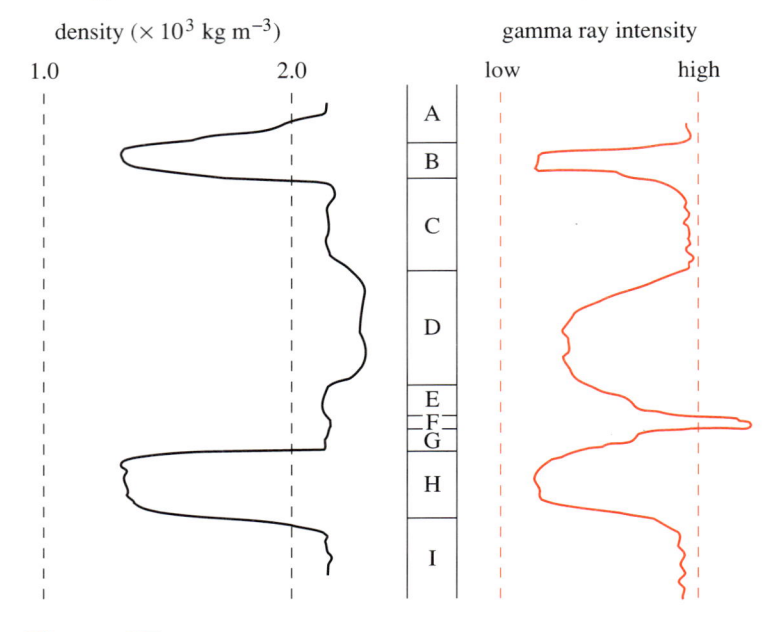

Figure 47 A geophysical log showing the density and gamma ray intensity data for a coal-bearing sequence.

Question 21

Explain the meaning of the term 'stripping ratio' in relation to opencast mining. Why can this ratio be greater for anthracite coals than for coal used in power stations?

Question 22

When a coal face first enters a new area of reserves it may encounter a variety of geological hazards, which may halt it. Which hazards may be described as (a) gradual changes, (b) sudden changes? What would be the effect on production of each of the two classes of hazards?

4 FINDING AND EXTRACTING PETROLEUM

Natural seepages of oil and gas have been known about for as long as coal. In 1272, Marco Polo described a sacred flame burning natural gas in the Zoroastrian fire-temple at Baku in Azerbaijan. In 1859 the world's first major underground oilfield was discovered 21 m beneath Titusville in Pennsylvania. The rapid rise of oil onto the energy scene was assured when, from 1870 to 1890, German engineers developed an engine that ran on fuel oil yet was light enough to power a moving road vehicle. Petrol engines changed the world in the century that followed.

This Section considers the techniques used to explore for petroleum and explains briefly how oil and gas are produced from underground fields. Alternative sources of petroleum are considered, together with the environmental impacts of producing oil and gas. The Section ends with an assessment of current oil and gas reserves.

4.1 Fluids and petroleum migration

Petroleum occurs naturally in three physical forms; solid, liquid and gas. In many physical aspects, solid petroleum is similar to coal, but liquid and gaseous petroleum are quite different. Crude oil and natural gas are fluids which are capable of flowing from one site to another within rocks, so exploration and production strategies for oil and gas are totally different from those we have examined for coal.

Movement of fluid petroleum through rocks is called **migration**. There are two main reasons why fluids migrate through rocks. The first is due to loading and compression. When source rocks are buried they are compressed by the weight of rock above them. The compressed sediment becomes denser and loses some of its intergrain porosity. Fluids, including any petroleum within the pore spaces between grains, are expelled in just the same way as water can be squeezed out from a sponge. Fluids move *down* the pressure gradient, which is generally *up* towards the surface.

The second reason relates to the physical effects of chemical processes that accompany hydrocarbon formation. During the early stages of petroleum formation, long-chain hydrocarbons separate out from kerogen. These chains occupy more space than the kerogen and therefore they increase the local fluid pressure within the source-rock pore spaces. Fluids are therefore expelled because of *excess pressure* within the source rock compared with the surrounding rocks.

Expulsion of oil and gas out of the source rock through source-rock pore pressure is known as *primary migration* (Figure 48). Fluids will migrate from areas of high pore pressure to areas of lower pore pressure, which might be either upward or downward through the adjacent rocks.

Once expelled from the source rock, buoyancy drives hydrocarbons from depth up towards the surface. This is known as *secondary migration*. As we saw in Block 3, Section 3, fluids move easily through rocks which have a high permeability. Just as the hydraulic conductivity (the ease with which an aquifer transmits water) depends on both the permeability and the physical properties of the fluid (density and viscosity, Block 3, Section 3.4), so the effectiveness of secondary migration through a rock sequence depends on the permeability of the rocks and the density and viscosity of the fluid petroleum flowing through them.

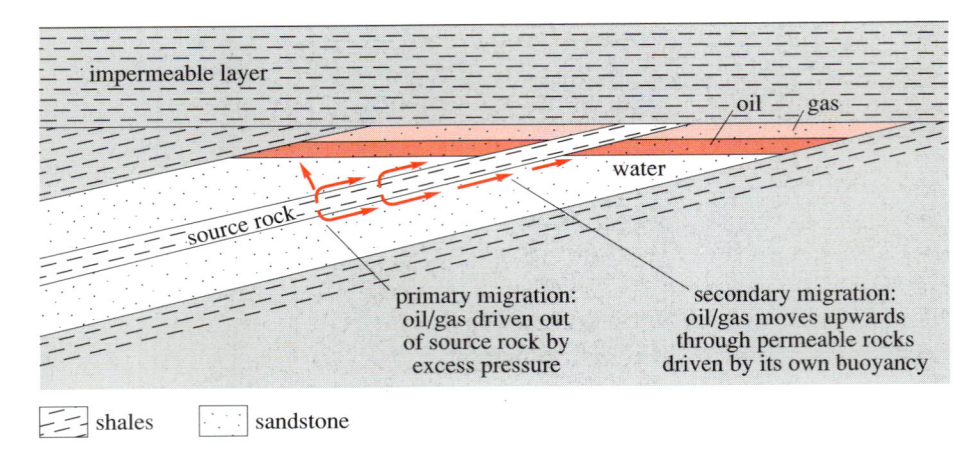

Figure 48 Migration of oil and gas out of a source rock and upwards through a reservoir.

Crude oil and natural gas are less dense than groundwater, so they move upwards through water-saturated rock. As Figure 48 shows, oil and gas continue to migrate upwards until they are trapped beneath some impermeable rock layer. At that point the hydrocarbons collect in layers according to their density: gas is lightest so it will pool at the top of the rock sequence; oil is heavier and will pool beneath the gas. Rocks beneath will be saturated with water.

4.2 Reservoirs, seals and traps — the play concept

The region around the southern part of the North Sea contains some of Europe's largest gas fields. The most substantial of these, near Groningen in Holland, initially contained 2.69×10^{12} m^3 of gas when it was discovered. The six fields in the UK sector of the southern North Sea named on Figure 49 together initially contained 2.17×10^{12} m^3 of gas, enough to supply the UK

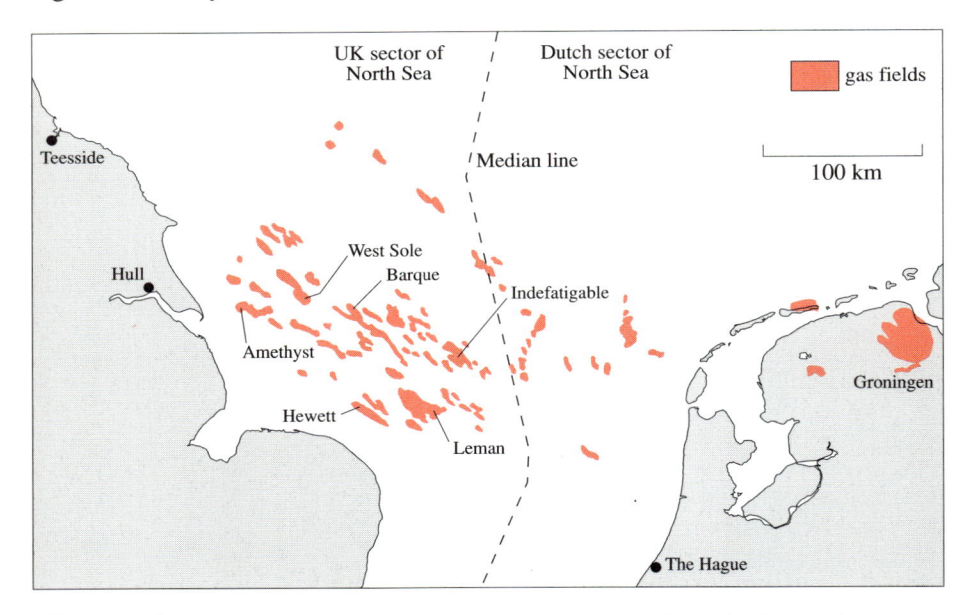

Figure 49 The location and extent of the principal gas fields in the southern part of the North Sea and the Netherlands. The Groningen field in Holland and the six largest gas fields in the UK sector of the southern North Sea are named. The median line marks the boundary between the offshore areas of UK and the Netherlands.

for 37 years at 1992 rates of use. The gas contained in all these fields came from burial and thermal maturation of Carboniferous coal seams.

Compare the locations of the main gas fields shown on Figure 49 with the extent of Carboniferous coal under the region shown in Figure 39. Coal lies at depth beneath all the major gas fields, but there are many parts of that area which do not contain gas fields. Why should that be?

The answer to this question contains the whole rationale behind petroleum exploration. Petroleum fluids are free to rise through the rock sequence, away from their source layers. So, unlike coal, it is not sufficient to locate source rocks at depth when exploring for petroleum.

The rock in which petroleum becomes ponded must, like an aquifer, contain enough interconnected pore spaces to store significant quantities of fluid. The impermeable barrier must contain no fluid pathways, otherwise fluids would escape. Furthermore, the barrier must have a substantial, roughly dome-like three-dimensional shape to trap large quantities of fluid. The major gas fields shown in Figure 49 have only formed where *all* these geological preconditions have been met.

There are four essential prerequisites for oil and gas accumulations (defined in the following sections):

1   The region must contain suitably mature *source rocks*.
2   These rocks must be connected by a permeable pathway to a significant layer of porous and permeable *reservoir rock*.
3   The reservoir must be covered by an impermeable *seal* or *cap rock* to prevent further migration.
4   The shape of the reservoir–seal interface must be such that it will *trap* a significant volume of hydrocarbons.

4.2.1 Reservoir rocks

The properties of a petroleum **reservoir rock** are very like those of an aquifer (Block 3). Petroleum can be contained both within the pore spaces, and also within fractures in the rock.

● Which of the three types of rock — igneous, sedimentary and metamorphic — will usually make the best reservoirs?

○ Almost all reservoir rocks are sedimentary rocks. The texture of igneous and metamorphic rocks, with close interlocking minerals, generally means that they will have low porosity and are almost always unsuitable.

Sedimentary rocks that are well cemented will have only small voids between grains and hence low porosity. The most porous reservoir rocks are generally well-sorted, poorly cemented sandstones, and these make up some of the most important petroleum reservoirs around the world.

Migrating waters can increase porosity and permeability on the scale of millimetres or centimetres by dissolving the cement that holds the grains together and widening small fractures that run through the rock. This effect is often enhanced if the waters are slightly acidic. Limestones are not naturally porous rocks because they are usually well cemented, but the calcium carbonate ($CaCO_3$) that makes up both grains and cement is soluble in acidic water. Consequently limestones can form good reservoirs, and in fact limestones hold 40% of the world's resources of petroleum.

Question 23

Which of the following geographical settings should give good reservoir rocks and which unsuitable reservoir rocks:

(a) a desert environment, like in the present-day Sahara desert in Africa, in which the wind constantly shifts sand, winnowing away fine material and leaving behind coarse material;

(b) a coral reef, like in the present-day Great Barrier Reef of Australia, made from limestone which itself results from the skeletons of a framework of corals and associated organisms;

(c) a delta, like the present-day Mississippi delta in America, where a major river enters the sea, depositing sand-size material and permitting mud-size material to be swept further out to sea.

4.2.2 Seals

The capping rock above a reservoir must be impermeable if it is to be effective in sealing petroleum into the reservoir. The porosity of a **seal** (also called cap rock) is not the critical factor — it may contain voids as long as the voids are not interconnected. Cap rocks are almost always sedimentary rocks that do not develop fractures, because they are composed of a material that responds to stresses within the Earth by flowing (to a limited extent) rather than breaking. Two rock types in particular have these properties; *shales* and *evaporites*. Other lithologies can form seals, such as the fine material that is produced along the surface of a rock fault by differential movement of the two sides. Shales seal about two-thirds of the world's oil and gas fields; evaporites seal most of the rest.

4.2.3 Traps

Oil and gas that accumulated as a thin layer at the top of an extensive horizontal reservoir would be uneconomic to extract. The seal–reservoir interface must be shaped to **trap** substantial quantities of hydrocarbons, rather like the curved upper surface of a balloon traps buoyant hot air. There are three principal types of structure that can trap oil and gas:

1. *Structural traps*, which are formed as a result of the deformation of strata in the Earth's lithosphere by folding or faulting.

2. *Stratigraphic traps*, which occur where there is a permeability barrier caused by lateral and/or vertical variation in rock types.

3. *Combination traps*, which are traps formed by both structural and stratigraphic means.

About 78% of the world's crude oil resources are held in structural traps, 13% in stratigraphic traps and 9% in combination traps.

Structural traps

The most common form of structural trap is an **anticline** or upfold within the rock layers. Anticlines account for the largest volume of the world's oil and gas resources. They are common in regions where the lithosphere has been compressed, and they can be large structures. In Figure 50, the petroleum is trapped because the upper parts of the reservoirs are now surrounded by a seal. In three dimensions the structure must be a *dome*, in which the rocks dip or slope downwards and outwards from the centre in all directions.

As oil fills the dome, water will be displaced out of the reservoir rocks. If both oil and gas are being trapped, the lighter gas will occupy the upper levels, floating on the oil beneath. The structure is full when oil and gas occupy the maximum volume within the dome. Any further introduction of oil or gas will

4 Finding and extracting petroleum

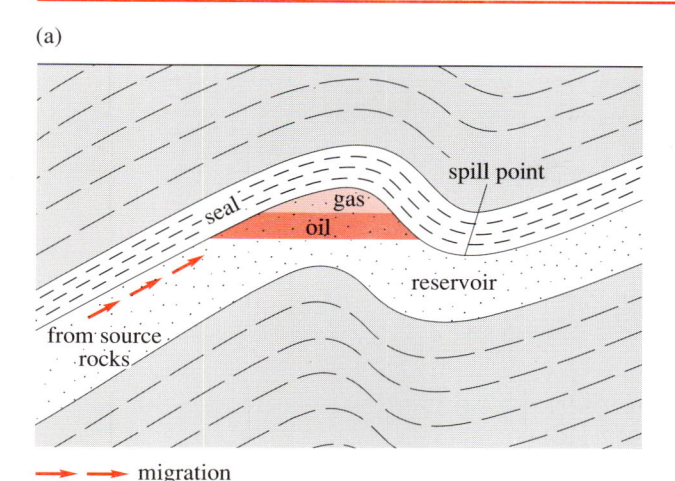

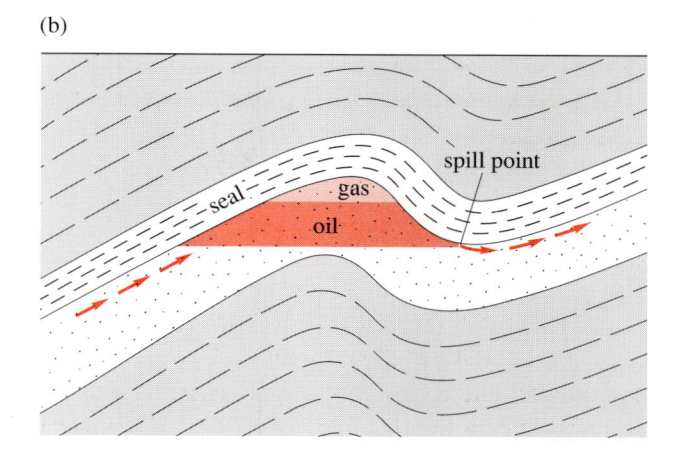

cause some oil to spill out from the structure at the **spill point** (Figure 50b). Spilt oil will continue to migrate upwards, possibly into another trap.

The volume of petroleum in an anticlinal trap is controlled amongst other factors by the lateral size of the structure (anticlines can be tens or, exceptionally, hundreds of kilometres long) and the amount of *closure* on the structure, i.e. the vertical distance between the position of the spill point and the top of the arched reservoir. Large anticlines can have hundreds of metres of closure.

Anticlinal structures can sometimes result from compaction of sedimentary rocks over a rigid block of older rock. This type of structure is illustrated in Figure 51.

Figure 50 The migration of oil and gas into and out of an anticlinal trap. (a) During the early phase of oil migration the lower level of the oil in the reservoir has not reached the spill point. (b) Migration has continued until the spill point has been reached by the bottom of the oil column, and oil can continue to flow further along the reservoir.

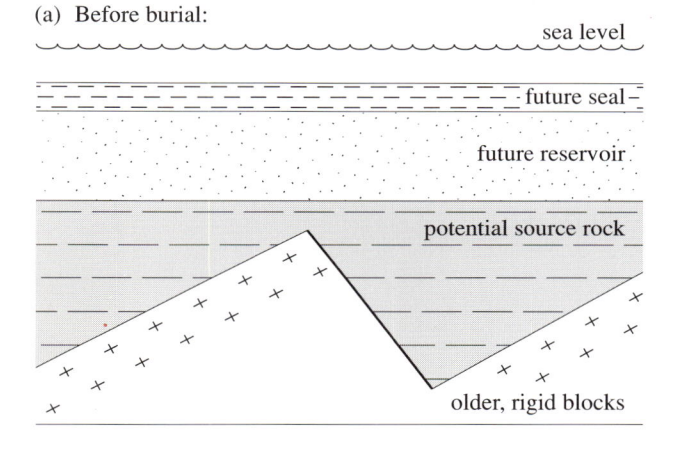

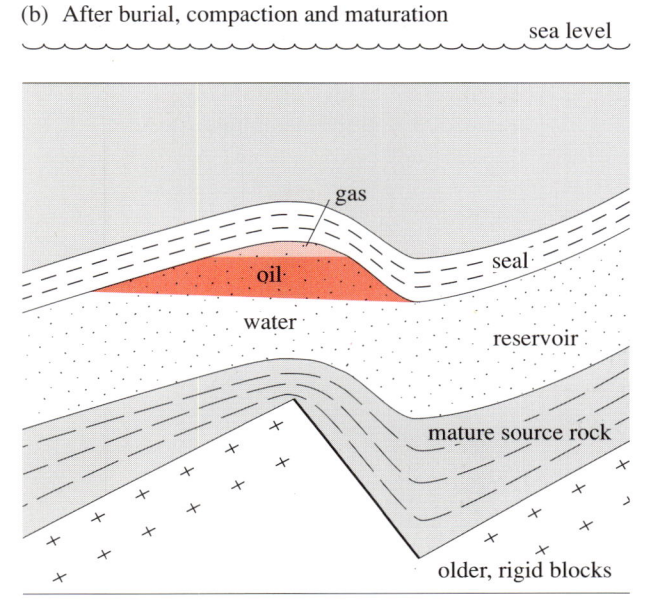

Figure 51 Schematic cross-sections (not to scale) showing how an anticline can form by compaction over rigid older blocks: (a) before burial; (b) after burial, compaction and maturation.

Other anticlinal structures can form in a broadly similar way by **salt movement**. These develop over sequences containing evaporites. When deeply buried and under high pressure, salt behaves in a plastic way. Its relatively low density means that when it flows, it tends to rise towards the surface, punching through the rocks above in elongated cylindrical plugs or pillars of salt, commonly about 1 km in diameter. Traps are formed as the plug pierces the overlying strata, which may by chance include suitable reservoir rocks (Figure 52). Some salt pillars can be up to 10 000 m high; they may pierce many reservoirs and form many traps.

Large blocks of reservoir and/or source rock are sometimes isolated within the salt plug during its formation. These blocks can contain oil and gas fields in their own right (Figure 52, locality A) and can be exploited if they are large enough.

Fault traps are formed when a bed of reservoir rock is brought into contact with impermeable strata by movement along a fault. Fault trap types are illustrated in Figure 53. Where the complete thickness of the reservoir has been faulted against a sealing rock there is said to be *complete closure*; in other cases the closure is *limited* and there are spill points.

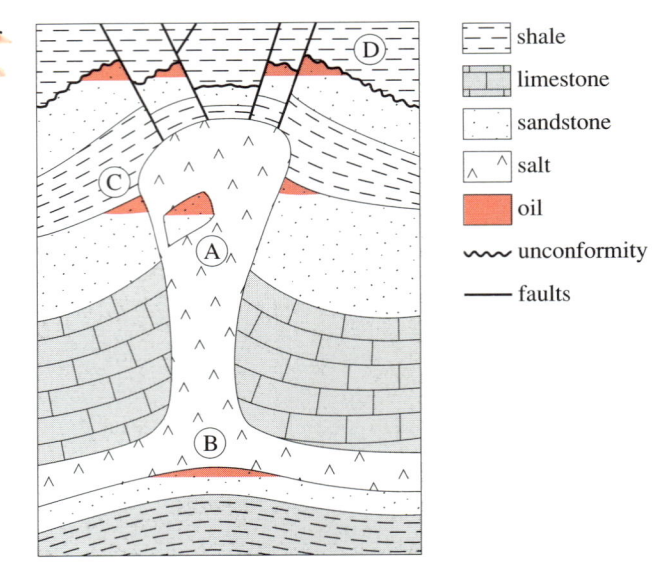

Figure 52 A schematic cross-section through a salt plug, with associated oil trapping possibilities. A, isolated block of source rock and reservoir wholly contained within salt plug. B, salt horizon seals an anticlinal trap. C, salt emplacement creates trap and salt forms part of the seal. D, structures in overlying rocks related to salt emplacement but not in contact with salt plug itself.

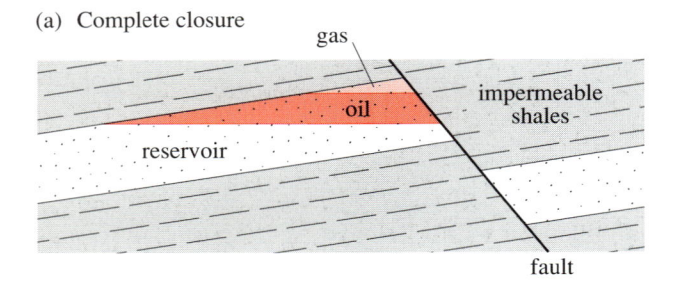

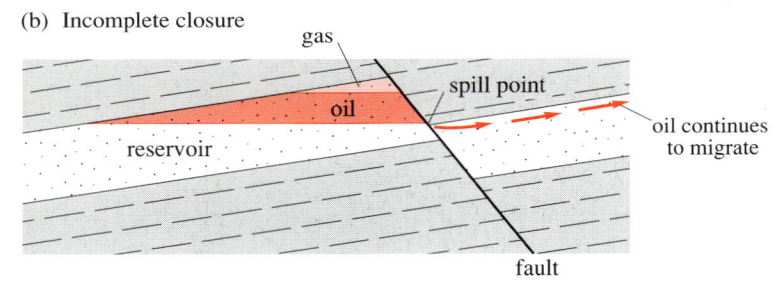

Figure 53 Fault trap showing (a) complete and (b) incomplete closure.

Question 24

Under the relatively low temperatures and pressures that characterize oil and gas basins, folding or faulting of rock sequences is usually a brittle process which leads to the development of fractures. In terms of oil and gas resources, is this a good or a bad thing?

Stratigraphic traps

Sediments vary in composition, grain size, shape and sorting as you trace them from place to place. From modern analogues, we know that a sandy bed deposited near shore, say, can change laterally to muds which were deposited in deeper water at the same time as the sandstone.

Some sandstones are discontinuous within shale deposits and appear like lenses (Figure 54a) or wedges (Figure 54b) in cross-section. Traps generated from lateral changes in rock type are called *stratigraphic traps*.

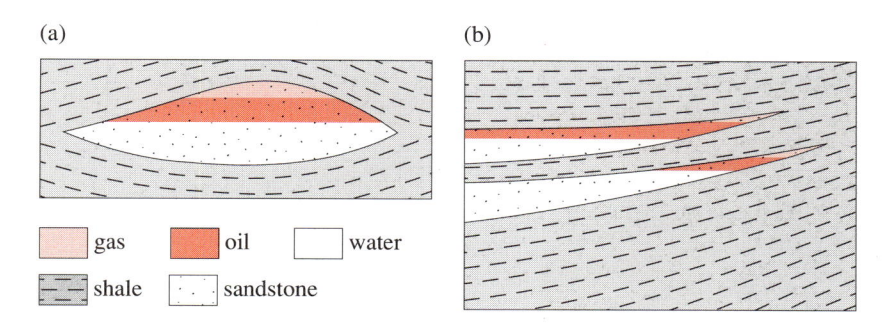

Figure 54 Two stratigraphic traps: (a) a lens-shaped sandstone body; (b) two wedge-shaped traps.

Reefs similar to modern coral reefs have existed in the geological past. Like some modern reefs, the fossil reefs are composed of limestone made up from the shells of marine organisms, which may later dissolve to yield a porous rock. When reefs become buried underneath impermeable sediment, such as mud, they can form sealed reservoirs known as *reef traps*. Such traps are restricted to rocks formed in former tropical and subtropical climates and account for only 3% of the world resources of petroleum.

The most important stratigraphic trap is the *unconformity trap*. An **unconformity** is a depositional break between two sequences of rocks of different ages. Frequently, unconformities have tectonic origins. If potential reservoir rocks are deposited, then tilted by tectonic forces, they may well subsequently be eroded, leading to a break in the rock sequence. If younger, impermeable rocks are then deposited on top of them a sealing unconformity will form.

In Figure 55 a period of uplift and erosion spanning some 70 million years has resulted in tilted reservoir beds below an unconformity overlain by nearly horizontal shales, which act as a seal. These reservoir rocks, in the Snorre Field in the Norwegian sector of the North Sea, are Triassic sandstones about 200 Ma old, sealed by Cretaceous shales, 130 Ma old.

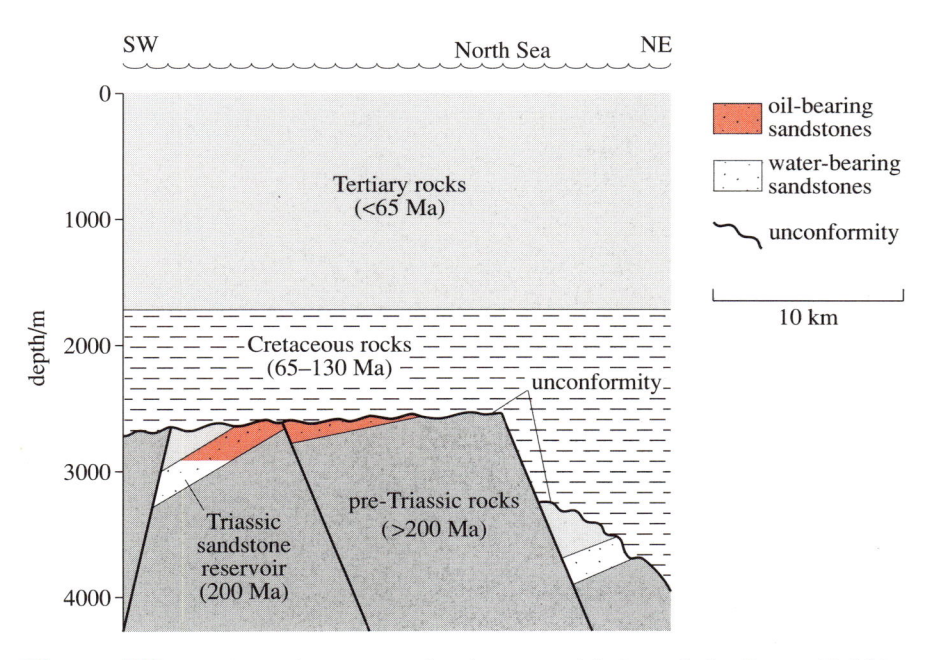

Figure 55 A schematic cross-section (not to scale) through the Snorre Field in the Norwegian sector of the North Sea. Oil is trapped beneath an unconformity where the difference in age of rocks on either side is over 70 million years (70 Ma).

Combination traps

These traps typically occur where reservoir rocks have been folded or faulted, and subsequently eroded and sealed by the deposition of shales on top of an unconformity. Such traps only contain petroleum if migration took place after the reservoir had been sealed by the much younger rocks. Figure 56 illustrates schematically a combination trap of the type found in the Prudhoe Bay field in northern Alaska. The trap is an east–west anticline which was tilted westwards and eroded. The overlying shales were deposited much later, and only after this seal was deposited was the whole structure buried deeply enough to generate petroleum from source rocks.

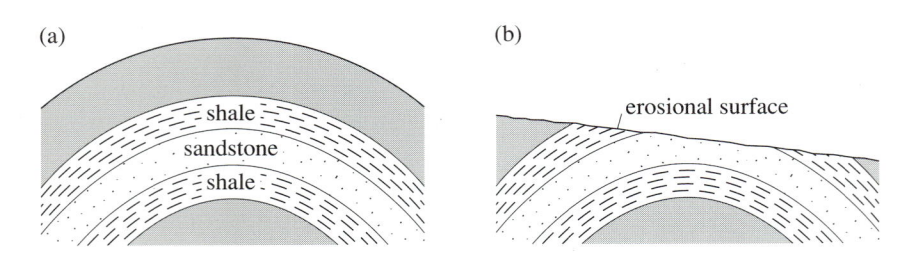

Figure 56 A combination trap. (a) Stage 1, the reservoir rocks are folded into a dome. (b) Stage 2, tilting followed by erosion exposes the reservoir rocks. (c) Stage 3, shale deposited on top of the unconformity acts as a seal and creates a trap.

4.2.4 The petroleum play

A **play** is an idea, a plan or a model of how mature source rocks, suitable porous and permeable reservoirs, adequate seals and appropriate traps have combined to produce accumulations of petroleum *at a particular stratigraphic level*. The petroleum play forms the basic strategy for oil and gas exploration. The play concept is perhaps best understood by illustration: read the Box about the Permian gas play in the southern part of the North Sea basin.

Play concepts are important because they aid oil and gas exploration on the broad scale, and the existence of a play in one area helps exploration geologists focus their thinking when they look elsewhere. Knowing that gas sourced from Carboniferous coals can be found in Lower Permian sandstones in the southern North Sea, for example, suggests that gas might be found in Lower Permian sandstones in onshore areas nearby.

- In that case, why has gas *not* been found in Lower Permian sands beneath the Chalk downs of East Anglia?

- No gas is found under East Anglia because neither the Lower Permian sands nor Carboniferous coals exist at depth beneath the Chalk of East Anglia (see Figure 39 to confirm the latter).

There is no reason why good source rocks, reservoirs, seals and traps should be continuous across an entire region, as the East Anglian example shows. The play concept alone is not enough to find oil and gas fields.

4 Finding and extracting petroleum

The Permian gas play in the southern North Sea

The story of oil and gas in the North Sea started with the discovery in 1959 of the huge Groningen gas field in The Netherlands. At the end of the 1950s, there was very little information about the rocks under the North Sea. The small size of the existing oil discoveries in east England, north Germany and the Netherlands did not offer much encouragement for companies to explore in harsher offshore conditions.

The best indications of solid geology were the outcrops to be found onshore, around the perimeter of the southern North Sea. Geologists knew that most of northern Germany, the Netherlands and northern Belgium were covered by thick Quaternary sediments. However, Cretaceous rocks were known to occur in northern France and southern Holland, as well as in south-east England, and Carboniferous rocks formed the major coalfields of western Germany, north-east France and central England. Rocks as old as the Lower Palaeozoic were known from the English Midlands and the Ardennes massif of southern Belgium. It was possible that the southern North Sea area contained an extensive basin of sedimentary rocks.

In the late 1950s, boreholes drilled in the Netherlands proved a thick Tertiary and Mesozoic sequence overlying Upper Palaeozoic rocks. In 1963 the Ten Boer-1 well near Groningen in the Netherlands (Figure 57) was deepened to around 3000 m to penetrate Lower Permian sandstones. This well discovered a giant gas field containing 2.69×10^{12} m^3 of recoverable gas, almost 2% of known world gas reserves.

A new play had been discovered. The Groningen field contained gas sourced from deeply buried Upper Carboniferous coals. The reservoir was a Lower Permian wind-blown sandstone with porosities ranging from 12% to 20%, sealed by thick Upper Permian salt horizons.

This play was exciting for British geologists because similar geology to the Groningen source, reservoir and seal could be seen onshore in Durham and Yorkshire. Detailed studies of the Groningen reservoir sandstones showed that they formed in a low-lying desert basin situated at about 20° N in Permian times. The Permian trade-winds blew from east to west. Was it possible that this desert extended offshore under the southern North Sea? The play as a whole could even extend from mainland Europe as far as Britain, and with it the prospect of other finds like Groningen.

The search for Groningen analogues offshore started in 1962 and drilling commenced a couple of years later. The first discovery to be made, in 1965, was the West Sole gas field which lies 380 km west of Groningen. The exploration borehole penetrated a 150 m thick Lower Permian desert sandstone reservoir at 2977 m, sealed by Upper Permian evaporites. The West Sole structure proved to be an elongated and faulted dome, 19 km long and 5 km wide, containing 50×10^9 m^3 of recoverable gas. The gas is almost pure methane, and geochemical data confirmed its origin from the Upper Carboniferous coals beneath. Together, West Sole and Groningen proved that the Permian play stretched across the whole of the southern North Sea.

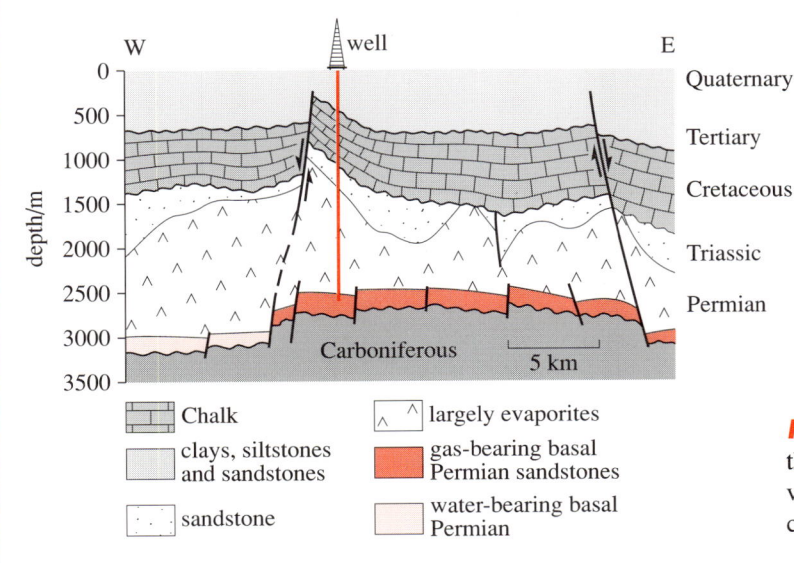

Figure 57 A cross-section through the Groningen gas field. Note that the vertical scale is greatly exaggerated compared with the horizontal scale.

4.3 Exploring for oil and gas

The surface geology of a region may not give any indications that there are petroleum plays beneath. Groningen is a good example; the Quaternary sands and clays that form the surface layers of Holland reveal nothing about the possible existence of Carboniferous coals and Permian sands at depth.

- Would a programme of borehole drilling have been an appropriate way to locate the Groningen gas field?

- A close-spaced pattern of boreholes across Holland would eventually have revealed the Groningen gas field, but the cost of repeated drilling to depths of around 3 km would have been colossal. Any profits that might have accrued from extracting and selling the gas would already have been wiped out by the high exploration costs.

Drilling is not the best way to explore for petroleum. Over the last seventy years or so, geophysical methods have been developed that allow the rock sequence below ground to be investigated without the need to drill expensive boreholes. Exploration for oil and gas is nowadays largely an exercise in remote geophysical analysis, directed towards answering two main questions. Could mature source rocks exist in this area? Are there suitably large traps, structural or stratigraphic, that could enclose economic volumes of oil or gas?

4.3.1 Regional geophysical exploration

Remote geophysical techniques used for exploration vary both in effectiveness and cost. Generally speaking, techniques which allow whole regions to be surveyed are relatively inexpensive but locate only broad areas which might contain oil or gas. Specific targets demand more detailed surveys. One use of regional geophysical studies is not actually to find oil and gas fields but instead to eliminate areas within a broader region in which the rock sequence is generally unsuitable for petroleum formation.

Gravity surveys locate sedimentary basins. Sedimentary rocks usually have a lower density than igneous or metamorphic rocks. The thick sequences of relatively low density sediments found in sedimentary basins effectively reduce the Earth's gravitational pull in that region, so substantial thicknesses of sediments can be located by measuring the Earth's gravity field and finding regional lows.

Regional gravity surveys can be made quickly and cheaply. Figure 58 shows a typical gravity anomaly over the thick sedimentary basin off central Wales in Cardigan Bay.

However, gravity lows are not diagnostic of sediments alone because other rocks can also show regional gravity lows. Granites are significantly less dense than other igneous rocks and approximately the same density as sediments. Large granite intrusions are also marked by regional gravity lows and in some circumstances can be indistinguishable from sedimentary basins.

Regional **geomagnetic surveys** can also be effective in locating sedimentary basins. Magnetic rocks cause perturbations in the Earth's magnetic field, whereas non-magnetic rocks have little effect. Investigating the Earth's magnetic field over any area can therefore be a guide to what rocks occur there. Sediments are typically poorly magnetic because they do not generally contain large amounts of iron-rich minerals. Other rock types, particularly volcanic lavas, are high in iron-bearing minerals and so are highly magnetic. Sedimentary basins are characteristically areas of uniform magnetism,

4 Finding and extracting petroleum

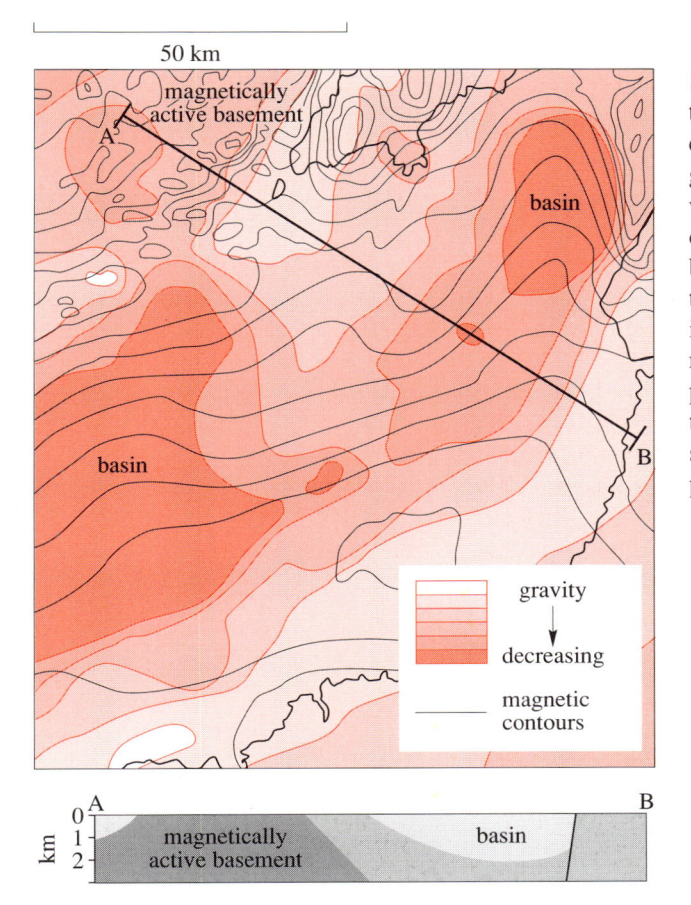

Figure 58 A map of the Cardigan Bay area off central Wales, showing gravity and magnetic values. The gravity low over a thick sedimentary basin is shown in colour tones. Magnetic contours indicate that old basement rocks, unlikely to be prospective in petroleum terms, lie close to the surface in the northern part of the map.

whereas volcanic rocks or old metamorphic 'basement' rocks (of little interest to the petroleum industry) near the surface typically show a highly variable magnetic structure.

Figure 58 also shows the uniform magnetic pattern over the Cardigan Bay sedimentary basin, together with the variable pattern over the Lleyn peninsula in the northern part of the map. The magnetics indicate that the Lleyn is made from rocks that affect the Earth's magnetic field strongly, so they are unlikely to contain oil and gas.

Both magnetometers and gravimeters, the instruments used to measure the Earth's magnetic and gravitational field respectively, can be carried aboard ship, by aircraft (not gravimeters) or even by satellites.

- From the *Postcard Geological Map*, what age and rock type are the rocks in the Lleyn peninsula? Why should they be highly magnetic? Why are they unlikely to contain oil?

- The rocks in the Lleyn peninsula are Lower Palaeozoic in age and contain large volumes of both volcanic and intrusive igneous rocks. They are highly magnetic because of the high proportion of iron-rich minerals in these old igneous rocks. They are unlikely to contain oil because lavas are unsuitable both as source rocks and as reservoirs, and because any sedimentary source rocks within very old rocks like these probably will already have lost any hydrocarbons they may have produced.

Regional surveys can locate sedimentary basins; they are good techniques to identify areas in which source and reservoir rocks may have been deposited and subsequently become mature. However, locating traps requires a much more detailed form of geophysical surveying, known as *seismic surveying*.

4.3.2 Seismic surveys

Seismic surveying is a geophysical technique used extensively by the oil industry to identify rock sequences and structures. The method is based on recording the time taken for sound waves to travel from a source at the Earth's surface down into the rocks, reflect off some distinctive rock boundary, and travel back to surface detectors. This information can then be processed to give an image of the geological structures at depth.

Seismic surveys can be conducted both on land and at sea. The land-based situation is shown schematically in Figure 59. There are three principal components in any seismic reflection survey:

1. An artificial source provides the sound waves. In the early days of seismic reflection surveying, a small charge of dynamite was detonated, shown schematically in Figure 59. Nowadays, more environmentally sensitive variants are normally used — such as a heavy pad vibrated hydraulically or electromagnetically. At sea, compressed air guns are fired just beneath the water surface. The place where the shot is fired is called the **shot point**. Each shot point is given a unique number so that it can be located on the processed seismic survey.

2. An array of detectors is spread out about 100 m apart. These sense the arrival of the reflected seismic waves at the surface. They are either fixed to the ground, or towed just below the water surface.

3. Mobile recording equipment is used to make a digital magnetic tape record of the electrical output from the detectors.

When a shot is fired, sound waves leave the source in all directions. Some travel down through the upper rock layers to reach a layer with distinctively different seismic properties. At this layer they may be reflected in roughly the same way that light reflects off a mirror. For this reason, such layers are called *reflectors*.

The reflected waves rebound and travel back to the detectors, reaching them at a different (usually later) time from any waves that have travelled there directly (Figure 59). Their exact time of travel will depend on the speed that sound travels through the rock, or in other words the **seismic velocity** of the rock layers they travel through. Other waves may pass through the first layer and travel on to a second or third prominent reflector. If these are eventually reflected back to the detectors, they will arrive later than waves reflected from upper horizons.

The detectors are therefore sensing 'bundles' of seismic waves arriving at different times because they have travelled by different routes through the rock sequence. When all the energy has arrived at the detectors and been recorded, the array is lifted up and repositioned and another shot fired, and so on. Computer processing allows the amalgamation of recordings from all the shot points, filtering out unwanted signals of various sorts. The final result is a cross-section called a **seismic section**. The section has a horizontal distance axis calibrated

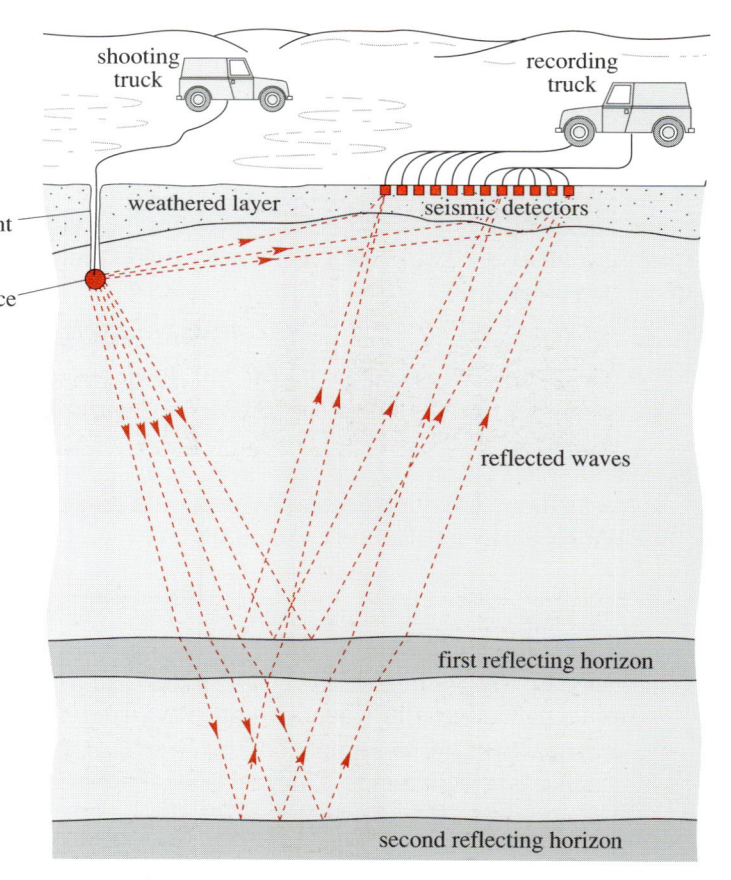

Figure 59 The layout of a seismic survey on land, showing the reflection (and refraction) of seismic waves by two reflecting horizons. The seismic waves are generated at the shot point and detected some distance away.

in shot-point numbers, which are spaced at regular intervals along the traverse line. The vertical time axis, calibrated in seconds, records the time interval from shot to arrival at the detectors, which in most cases is the **two-way travel time (TWT)** from shot to reflectors and back to the detector array.

Seismic sections show a series of light and dark lines. The intensity of the lines is related to the amount of sound energy reflected from different layers. Some rock types act as sonic mirrors, reflecting seismic waves strongly; others are almost transparent to sound waves. Strong reflectors are marked by intense black or white lines, contrasting with the weaker lines produced by other types of less reflective sediment junctions.

Figure 60 shows two typical seismic sections through the giant Brent oil and gas field in the northern North Sea, shot over the same structure twenty years apart. Figure 60a shows the quality of seismic section that was available to exploration groups in the late 1960s, when the first oil discoveries were being made in the North Sea. Figure 60b shows the vast improvement in resolution and detail that had taken place during the succeeding twenty years.

Both figures show the same rock sequence: horizontal or almost horizontal strata in the upper part of the section, separated from a sequence of gently dipping rocks in the lower part. The upper and lower sequences are separated by a plane of erosion, or unconformity. The ages of the two sequences are unknown from this section, but drilling has subsequently proved that the sequence above the unconformity is Cretaceous and Tertiary strata, whereas the lower, dipping sequence is Triassic and Jurassic in age.

Figure 60 Two seismic sections shot over the Brent oilfield in the Northern North Sea. Line (a) was shot twenty years before line (b). Line (a) shows the quality of seismic sections when Brent was discovered in 1971. Line (b) shows the vast improvement in quality that resulted from twenty years of technological advancement and knowledge of the seismic characteristics of the rock sequence in the Brent area. Note that the vertical scale starts at 2 s two-way travel time (TWT). (©1991 The Geological Society)

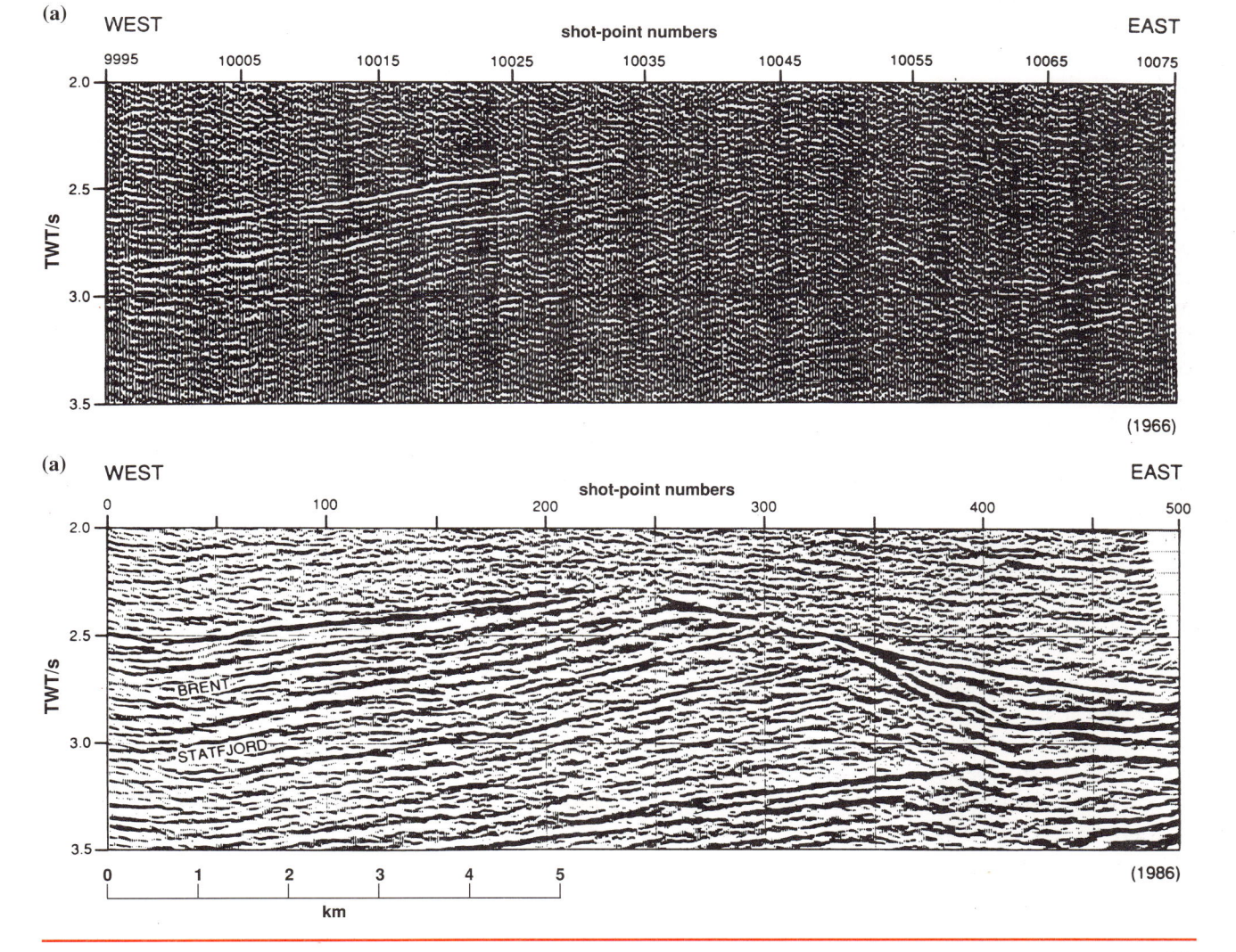

Block 4

Question 25

Unconformities are ~~characterized~~ *indicated* by changes of dip of rock sequences and by erosion of the lower, older sequence. By using these two features, identify and mark the line of unconformity on Figure 60b. (Since this is a difficult, rather subjective task, you could try starting at the point SP 260, TWT 2.35 s, which lies on the major unconformity, and trace the unconformity to the left and right from there.)

Interpreting seismic sections is a complex process, requiring both experience and a certain amount of interpretative 'flair', as you probably discovered attempting Question 25. However, it is a relatively cheap way of obtaining a continuous cross-section through a sequence of unexposed rocks. Usually a grid of sections will be shot, processed and interpreted by geologists.

4.3.3 Exploration drilling

When a suitable structure has been identified from seismic sections, the next step in the exploration process is to drill into the target sequence. The main objective of an exploration borehole is to establish if petroleum is in fact

---

Oil in western Canada in the nineteenth century

Historically, oil was only discovered where it seeped from the ground. Surface seepages of oil or gas led early prospectors to surmise that petroleum was stored underground and might be located by drilling near the source of the seeps. In fact, the majority of oil or gas fields discovered in the late nineteenth and early twentieth centuries were found from seeps. Sometimes, as was the case in Canada in 1890, oil or gas was actually found by prospectors who were drilling for coal.

Although oil seeps were uncommon, they had been noted by native Canadians, who passed the information to these early surveyors. Oil was frequently described in association either with dark oil shales, or with salt springs and it gradually became clear that the Canada was rich in oil. By 1874, Alfred R. C. Selwyn, the second Director of the Geological Survey of Canada, reported: 'There seems but little doubt that Canada has her a salt and oil bearing region surpassing in extent and productive capacity any hitherto developed on the American continent.'

Exploratory drilling started in the 1870s, spurred on by Selwyn's disappointment at the lack of outcrops in the Canadian plains. The construction of the Canadian Pacific Railway across the Canadian Prairies in 1882–83 led to drilling for water and, to a lesser extent, coal. The first of two wells drilled by the line of the railway in 1883 encountered, quite by chance, a heavy flow of gas from Upper Cretaceous sands at 323 m. The well had reached a depth of 352 m when the gas ignited and destroyed the drilling derrick. Gas from this well was used to fire the boilers for a second attempt the following year, and gas from both wells was used by

Figure 61 The Canadian-type pole drilling rig that first discovered oil in western Canada in 1902. This well reached 311 m. Note the wooden poles used for drilling at that time. The barrels in the background were used for storing crude oil.

the railway company for many years subsequently. Quite by chance, the drillers had discovered the Alderson gas field, with initial reserves of $13.43 \times 10^9 \, m^3$ of gas.

The first underground oil discovery in western Canada was made in Alberta in 1902. One John Lineham sunk a well to 311 m and found oil in Lower Cretaceous sands (Figure 61). The well yielded $47.7 \, m^3$ of crude oil per day. Oil had been discovered in one of the major sedimentary basins in North America. Almost a century later, oil and gas reserves in the whole basin prior to production were estimated at $2.86 \times 10^9 \, m^3$ of oil, $0.73 \times 10^9 \, m^3$ of gas liquids and $3.73 \times 10^{12} \, m^3$ of gas.

trapped in the target structure and, if so, how much is present in terms of thickness of reservoir rock containing oil or gas. Other important objectives include establishing the age, thickness, porosity and permeability of any potential reservoir. These can be important even if the exploration well turns out to be a *dry hole* which contains only water ('dry' for petroleum) at the chosen drill site.

Drilling for oil and gas is sophisticated process; these boreholes are deep, often over 3000 m and the deepest exceed 6500 m. At such depths the fluid pressure within the rock sequence is very high, so the drill bit is lubricated by using a drilling mud based on the dense barium sulphate mineral, *barite*. This dense mud maintains the pressure downhole and also carries cuttings of the rock sequence back to the surface.

Most of the drilling is completed using a solid bit; cores are generally only taken when petroleum-bearing reservoirs are met. A sophisticated array of geological and geophysical tools is used to take measurements whenever the drill string is removed from the borehole (to exchange a worn bit, for example) and when drilling has been completed. The details of these techniques are beyond the scope of this course, but remote measurements of *physical properties leading to estimates* rock type, porosity, permeability, dip and oil or gas content can all be made without necessarily recovering any of the rock. For horizons of particular interest, remote tools can be used to take samples from the walls of the borehole without the need to core.

4.4 Petroleum production

Once quantities of oil or gas have been discovered during exploration drilling, the next step is to carry out an **appraisal** of the field to develop a production strategy for the field as a whole. To move into production, the field must contain enough oil and gas to be able to repay the huge cost of development and day-to-day operation and still produce a profit.

4.4.1 Appraisal

During the appraisal stage, the size of the field must be established as accurately as possible and the most effective way to produce the highest percentage of petroleum from the reservoir(s) sought. Production geologists assess the details of the reservoir in particular. If the reservoir is extensively affected by faults, or if certain sections of the reservoir have poor lateral or vertical permeabilities, production could be adversely affected. The exact pattern of production wells needs to take such factors into account.

Nowadays, a **three-dimensional (3-D) seismic survey** is shot before appraisal is completed. This is no different in principle from the exploration seismic survey but is shot on a closely spaced grid (50 m centres or less between shot points) over the detailed area of the field. 3-D seismics gives much more detail about the rock structure, the depth to reservoir units and, to a certain extent, the nature of the reservoir rocks than was achieved during the exploration survey. Further drilling will improve the knowledge of rock types, distribution, permeability, temperature, pressure, and hydrocarbons content.

Production geologists frequently try to *model* the depositional environment of the reservoir by comparing the rock samples obtained from the borehole with modern analogues. If the most porous and permeable part of the reservoir was deposited as a beach sand, say, then the extent and nature of the buried sand body might be estimated by comparing the downhole samples with modern beach deposits. The depositional model may indicate areas in which permeability could be expected to be low, such as a muddy lagoonal area behind the fossil beach. Whether or not these deposits are preserved underground can be assessed during appraisal drilling.

Economic and political factors always have to be considered. Will the price of oil rise or fall over the lifetime of the field? Are new tax laws likely? How much will it cost to install production facilities, and to keep them efficient? How will the oil be moved to the refinery? Production will only start when all these questions have been answered satisfactorily.

4.4.2 Production techniques

All fluid petroleum is held underground at high pressure. During the early stages of petroleum production, getting these fluids to the surface safely means allowing a controlled escape of fluids under pressure. Later in the life of an oil or gas field, it usually becomes necessary to maintain the pressure underground by injecting pressurized water or gas, or both, into the reservoir (Figure 62).

The percentage of petroleum, particularly heavy oils, that can actually be recovered from a reservoir depends on the viscosity of the fluids. Thick, waxy oils are more difficult to extract than thin oils. Much can be left behind if the displacing fluids follow discrete pathways rather than flushing the oil out uniformly.

When production begins, during **primary recovery**, fluids within the reservoir begin to rise up the borehole and reach the surface. As the pressure is released, any gas dissolved in the oil will come out of solution, rise and escape along with the oil. As production continues, the pressure of the petroleum remaining in the reservoir begins to fall. The fall in pressure and the loss of dissolved gas cause increases in both the surface tension and the viscosity of the oil. It becomes 'thicker' and will not flow so readily. At best only 30–40% of the petroleum in the reservoir will be brought to the surface during the primary recovery stage.

When the natural drive of the petroleum dwindles, **secondary recovery** injection techniques are needed to increase recovery to about 50%. These techniques maintain reservoir pressure by injecting natural gas into the reservoir *above* the oil forcing the oil downwards, and flooding with water *below* the oil, forcing it upwards (Figure 62).

In order to improve recovery still further, new techniques have been developed, known as tertiary or **enhanced recovery** methods. These techniques include the addition of detergents to the injected water in order to reduce the viscosity of the crude oil, or the injection of steam. This can result in up to 90% of the initial oil being recovered, but detergents are expensive,

Figure 62 Oil and gas production techniques. When the natural pressure within the reservoir has dissipated, the drive can be maintained by injecting (a) gas into the top of the reservoir, or (b) water into the reservoir beneath the oil, or both.

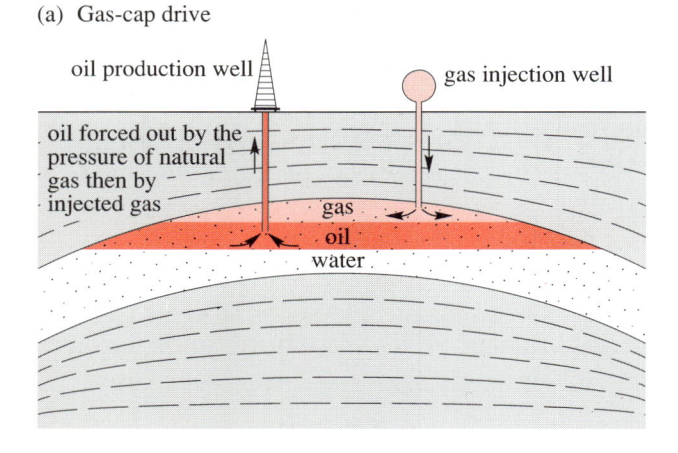

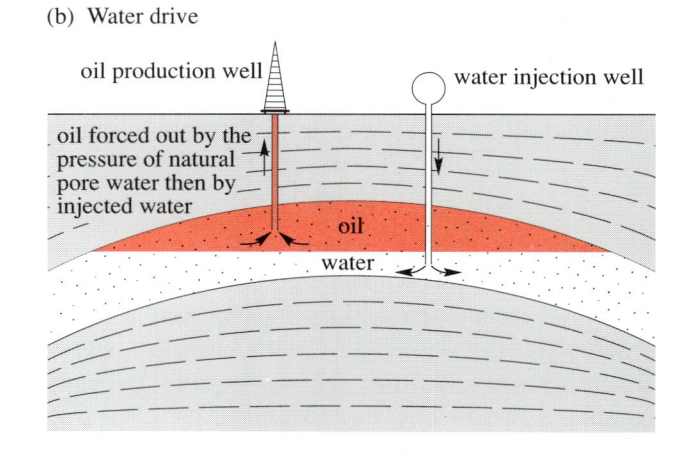

(a) Gas-cap drive

(b) Water drive

4 Finding and extracting petroleum

and for many smaller fields the amount of extra oil recovered may not be worth the investment. Even with modern recovery techniques, about 50% of the initial oil in place in any reservoir often has to be left there.

Now watch Video Band 11: *Energy Resources — Petroleum*, which deals with the techniques of oilfield appraisal and petroleum recovery. Check your understanding of its contents after you have watched it by answering Questions 26 and 27.

Video Band 11 Energy Resources — Petroleum

Speakers

Jim Dolph	Gulf Canada Ltd.
Grant Mossop	Alberta Research Council
Norman Wardlaw	University of Alberta
John Wright	The Open University

This programme was made in 1983. The video has three main aims:

(a) It examines the way in which lateral changes in sedimentary rock type (called *facies changes*) control the porosity and permeability of rocks. In turn, these control the way in which petroleum moves through the rock sequence.

(b) It illustrates how the Goose River oilfield in Alberta, Canada, was discovered by a combination of geological studies, geophysics and drilling.

(c) It illustrates with models how primary recovery using solution gas and gas-cap drives can yield about 30% of the contained petroleum, secondary recovery by water flood can yield up to 40%, and enhanced recovery using detergents can yield up to a total of 90% of the initial oil in place.

The following points made in the video are particularly important:

1. Modern sediments (in this case in East Anglia) can be used to show the kinds of lateral and vertical changes that occur in reservoirs underground, where they can only be accessed by drilling.

2. Jim Dolph summarizes the stages in the discovery of the Devonian oilfield at Goose River, north-west Alberta. This oilfield lies in a reef complex that lay to the south-east of an ancient shore line. The rocks dip gently to the south-west, and the oil–water interface runs approximately north-west to south-east. The oil lies updip, to the north-east of the oil–water interface.

3. Norman Wardlaw uses the champagne bottle analogy to illustrate how the initial gusher in an oilfield is produced by solution gas drive. Gas-cap drive then takes over as the gas cap expands in response to falling pressure, and drives the oil up the well pipe. Primary recovery rarely brings up more than 30% of the oil in any field.

4. A laboratory model is used to simulate secondary recovery by water flood processes. The oil is vigorously flooded out rather than sedately floating up above a rising oil–water interface.

5. Much oil is left behind in large pockets that have been by-passed by the water. Detergents in the water can break up and remove these by lowering the surface tension between the oil and the water. This can result in almost all the initial oil being recovered, but detergents are expensive so their application needs to be carefully controlled.

6. For effective tertiary recovery, the geometry of the reservoir has to be known in great detail. Jim Dolph describes how a detailed geological model was constructed using old drill cores. Comparison of the drill cores with models of modern reef environments enables Goose River to be identified as an ancient reef complex. This improved understanding allows more oil to be recovered, but is not expected to extend the life of the field beyond 2020.

7. Some 400 km to the east of Goose River lie the Cretaceous Athabasca tar sands (also see Section 4.7.1), on the edge of the large Devonian basin. Grant Mossop explains how the oil in Devonian source rocks has migrated updip towards the edge of the basin and into the overlying sandstones. The sands are now exposed at the surface and many of the volatile petroleum fractions have evaporated. The residue is a viscous tar which 'sweats' out of the sands on hot days. Hundreds of tonnes of sands are dug out daily by huge excavators, and treated with steam and hot water to separate oil from sand and pebbles. The cleaned oil is sent to refineries.

8. Oil reserves in the Athabasca deposit are huge, roughly equivalent to the total conventional oil in the Middle East. The recovery rate is much higher than for most underground oilfields.

Question 26

In the Goose River oilfield:

(a) What is the main reservoir rock type?

(b) What use was made of seismic sections in the field's discovery and evaluation, and why?

(c) Describe the processes that have been, and will be, used to produce oil.

(d) What percentage of the total oil in the reservoir will be produced during each of the primary, secondary and tertiary recovery stages?

Question 27

How did geological modelling enhance Gulf Canada's understanding of the Goose River oilfield?

4.5 A modern oilfield case study—the Brent field in the northern North Sea

The giant Brent oil and gas field is situated in the northern North Sea, some 190 km north-east of the Shetland Isles. The North Sea here is just over 140 m deep. The field was discovered in 1971 by the Shell/Esso joint venture, and is being developed by Shell UK Exploration and Production, based at Aberdeen. The discovery of Brent opened up the principal oil-producing area of the UK.

The first seismic sections were shot over the area in 1966. The discovery well was drilled in May 1971 and struck oil in Jurassic reservoir sandstones. In June 1972, a second appraisal well was drilled about 5 km south-south-east of the discovery well, which penetrated a column of 213 m of oil and 52 m of gas. Tests carried out on that well measured a flow of 6600 BOPD.* The fourth appraisal well, also drilled in 1973 discovered more hydrocarbons in the deeper Statfjord reservoir sandstones.

Brent came into production in November 1976 and is expected to remain in production until 2018. By that time it is hoped that more than 271×10^6 tonnes of oil, 157×10^9 m^3 of gas and 46×10^6 tonnes of natural gas liquids will have been recovered from the four production platforms.

4.5.1 Geology

Brent is typical of the Jurassic play in the northern North Sea. The hydrocarbons in the Brent field are sourced from Kimmeridge Formation rocks.

There are two migration routes from Kimmeridge rocks into the Brent sandstones: up the steep fault scarps to the east, and up the gentler dip slopes to the west. Primary migration took place through kerogen layers in the source rock; secondary migration was through both microscopic and larger fractures and thin sandy layers. Coals and organic-rich shales within the reservoir sequence have also sourced gas into the Brent field. The source rocks were mature and the Brent structure filled by early Tertiary times, some 60 Ma ago.

Oil and gas is trapped within two main reservoir units (Figure 63a). The upper (Brent Group) reservoir is some 260 m in thickness. The reservoir units are all fine- to coarse-grained well-sorted sandstones, ranging in porosity from 16% to 29%. Thin shales and coals at discrete horizons within the reservoir sequence act as barriers to upwards fluid movement. The lower (Statfjord

---

* The oil industry takes its lead from America, and oil or gas volumes and rates are more usually quoted in **barrels**, or barrels of oil per day (**BOPD**) than in metric or SI units. One barrel holds about 0.14 tonnes of oil.

4 Finding and extracting petroleum

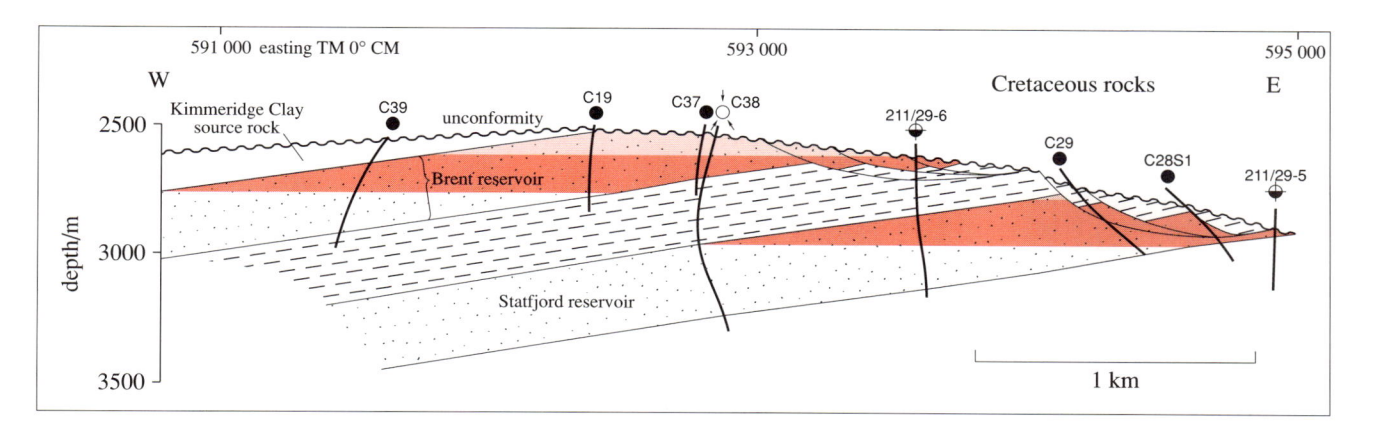

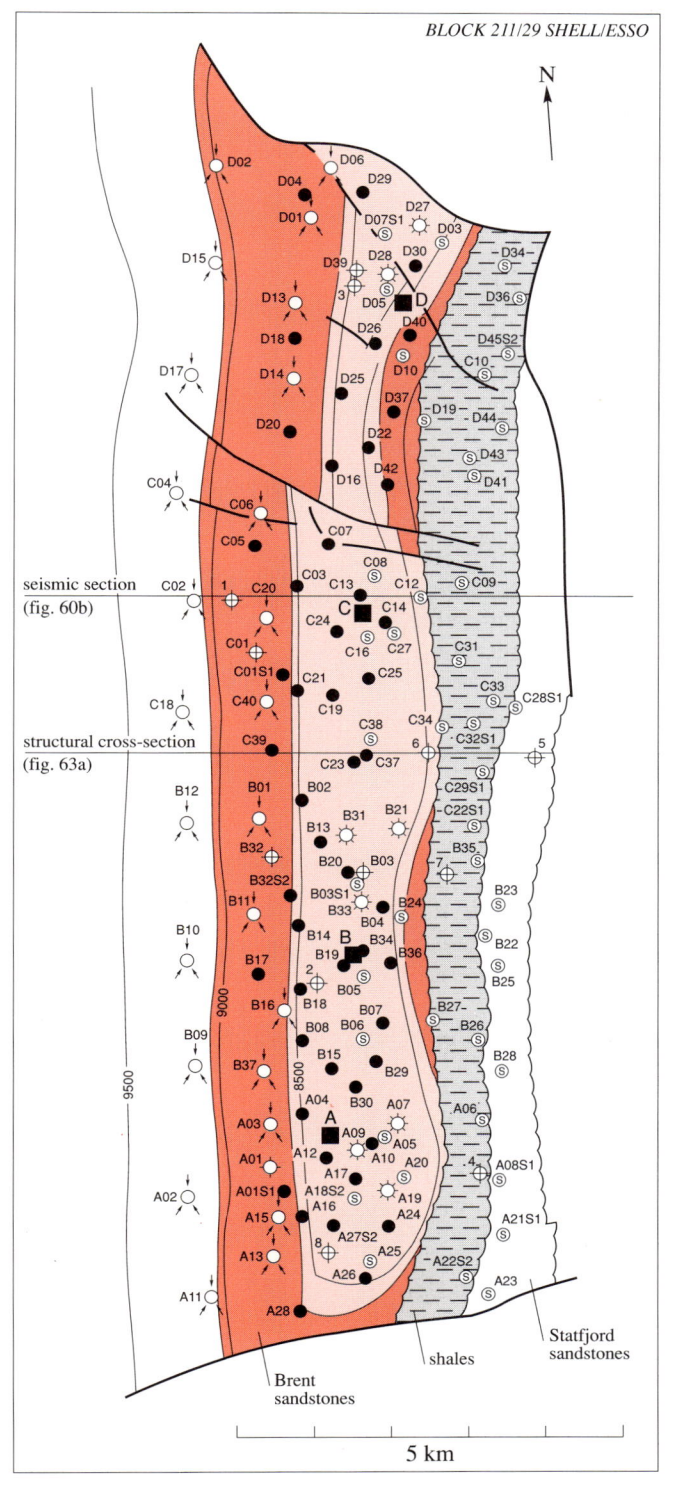

Figure 63 (a, above) Cross-section and (b, left) map through the Brent field, showing the shape and size of the field and the wells that have been drilled to produce its oil and gas. Well C38 (near centre of cross-section) is now a water injector. The symbols are the same in both diagrams. Note that the vertical scale in the cross-section covers only the depth at which the field is found: 2500–3500 m, and that the depth contours in (b) are in feet not metres (as is commonly the case in the American-dominated oil industry). These two diagrams are simplifications of a complex geological situation. They are included to show the variety of well types, locations and horizontal deviations that can be achieved from four production platforms. (©1991 The Geological Society)

Formation) reservoir sequence is a complex alternation of shales and sandstones.

The reservoir sandstones are overlain by both Kimmeridgian clays and mudrocks and Lower Cretaceous shales, which seal the field. These are overlain in turn by over 2500 m of Upper Cretaceous and Tertiary rocks. It was the deposition of these rocks that led to the source rocks becoming mature for oil and gas generation in Tertiary times.

The trap is a simple tilted fault block, dipping 8° to the west. The reservoir sandstones have been eroded and sealed by shales lying above an unconformity. Brent is, therefore, a combined structural/unconformity trap (Figure 63).

4.5.2 Geophysics

In the 1960s and early 1970s, wide-spaced rectangular grids of seismic lines, oriented east–west and north–south, were shot. Line spacing varied from 3 km to 7 km. These lines were processed before there was much information about the depths of the component rock units and their seismic properties, which only came after several wells had been drilled.

Between 1980 and 1982, a close-spaced grid oriented east–west (and north–south) was shot, to assist the production team with their modelling. The lines were shot every 200 m in a north–south direction, and with a shot-point interval of only 25 m. This closely spaced grid of data was then processed to give a 3-D data set giving a complete three-dimensional picture of the field. In 1986, a further 3-D grid was shot.

4.5.3 Production strategy

The initial field production plan was to develop both the oil zone and the gas cap of each of the Brent and Statfjord reservoirs. This would be achieved by waterflood and gas injection. Waterflood would maintain pressure within the oil-bearing part of the reservoir and displace oil towards the producing wells. Oil-producing wells would be re-sited when they eventually began to produce water rather than oil. At the same time gas injection would help to maintain pressure and store gas that had been released during oil production but was surplus to sales commitments. Using these secondary recovery techniques together, peak production of some 64 000 m^3 of oil per day was reached in 1986.

The oil is piped via the Brent system pipeline to the oil terminal at Sullom Voe in Shetland. Gas production varies, depending on the amount released during oil production. Fourteen million cubic metres per day is transported via a pipeline to St Fergus in Scotland and sold to British Gas. Any gas produced which is surplus to this is reinjected, and gas is also reinjected in the summer months when demand for domestic heating fuel is less.

In the early 1990s, Brent was providing 13% of the UK's oil needs and 10% of UK's gas requirements, but under the original waterflood production plan, the field would produce oil only until 2004. So in 1993 the Shell/Esso consortium replanned the Brent field production strategy to increase its expected life. The reservoir pressure has been dropped, allowing the dissolved gas to be released underground rather than at the surface. This gas can be produced from the wells on the crest of the structure, but it also helps to flush oil downwards into the oil-producing wells. This development should extend the field's life by at least six years to 2010, but Brent may continue to produce both oil and gas until 2015 if the new strategy works well.

As a direct consequence of this pressure drop, the sea bed is expected to subside by about 4 m. This has meant substantial redesign of the platforms and sea-bed installations throughout the Brent field, which in turn has meant increased capital expenditure. Ultimately Brent's life expectancy will be controlled not by geological factors but by the financial cut-off point, when it costs more to produce hydrocarbons than the price they can be sold for. When that comes, up to 40% of the original oil could be left behind, trapped by surface tension in the pore space between sand grains in the reservoir.

4.6 Environmental aspects of petroleum exploitation

Petroleum production, other than mining oil sands or shales, does not give rise to the same environmental hazard as coal mining. Only hydrocarbons fluids are produced from underground, so no waste or spoil heaps are generated. Workers do not go underground to produce oil, so there is no risk to health analogous to that found in deep coal mines, but offshore workers, particularly divers, have high-risk occupations.

Pressure drops when petroleum is removed during production, so minor subsidence often happens when a field is reaching the later stages of its life. In the Ekofisk field in the Norwegian part of the North Sea, the sea bed sank through compaction of the weak Chalk reservoir. This could have had serious consequences had it been an onshore site. Since 1985, subsidence in Ekofisk has been reduced by injection of gas and water. We have already discussed the expected subsidence attendant on changed production methods in the Brent field.

The most significant environmental risks from petroleum production come from spillages, either accidental or intentional. Spillages are rarely allowed to happen on rigs: there is such a great fire risk. All modern drilling rigs are fitted with control valves to prevent oil or gas 'blowouts'. Nevertheless, rig accidents do occur, like the tragic incident of the Piper Alpha platform in the North Sea where, in July 1988, 167 men died as the result of an explosion following a massive leak from the rig's gas storage system.

The most common deliberate 'spillage' from rigs used to be the flaring of gas. In fields where gas injection was not necessary and where no containment or transport facilities existed, gas was burned on site. This practice released large amounts of methane and carbon dioxide into the atmosphere. Fortunately, the incidence of gas flaring has decreased markedly in recent years, and a wasteful and environmentally damaging practice has largely been stopped.

4.6.1 The environmental impact of transporting oil and gas

Accidental petroleum spills are much more common as a result of oil transport. The immense size of modern sea-going oil tankers means that freak weather, poor charts or negligent seamanship can cause major oil spills. Most countries with significant coastlines now have disaster plans to cope with oil spills, but of course they do not always work. The relatively small spill of 37 000 tonnes of crude oil from the tanker *Exxon Valdez* in 1989 onto the coast of Alaska caused major environmental damage, but more than three times that amount of oil, 120 000 tonnes, was spilt into the English Channel in 1967 when the supertanker *Torrey Canyon* ran aground. However, environmental disasters are not inevitable. For instance, 85 000 tonnes of Norwegian light crude was spilled from the tanker *Braer* when it ran onto rocks at the south end of the Shetland Isles in January 1993. Fortunately, that oil was dispersed by a week-long hurricane. A year later, no trace of the oil remained at the site of the accident.

About 12% of all marine pollution is caused by oil transportation, but the incidence of major accidental spills (over 700 tonnes) is decreasing. In the 1970s, an average of 25 spills of this magnitude happened each year. Throughout the 1980s it had been reduced to nine spills every year on average, and in 1991 there were only seven such spills. Nevertheless, many analysts contend that more oil is 'spilt' each year by the deliberate flushing of tanks at sea than is lost by accident.

There is also a contribution to environmental damage caused by pipelines. Hydrocarbons can leak from faulty joints in pipelines, and even where spills are not a factor, pipelines transmitting warm oil through tundra or permafrost regions of the globe can damage the delicate ecosystem of those regions. The major pipeline from the north coast to the west coast of Alaska was resisted by environmental groups and had to be engineered to the strictest specifications to minimize environmental damage.

Obviously it is the intention of all oil and gas producers to minimize loss through accident or spillage, if only for economic reasons. Accidents, when they occur, are catastrophes for the areas concerned, and for the whole ecosystem when rare species are threatened. However, the lasting and permanent damage done to the environment by petroleum is not primarily through accidental causes but through our deliberate use of them as fuels, as we shall see in Section 5.

4.7 Unconventional sources of petroleum

4.7.1 Solid petroleum

When crude oil reaches the surface, the volatiles contained within it evaporate and the surface rock becomes impregnated with solid petroleum. If the surface rock has good reservoir properties, large volumes of oil can flow into it from mature source rocks below, but over time, exposure to air and bacteria will degrade that oil into thick, viscous tar or bitumen. This type of deposit is known as an **oil sand** or *tar sand*. The Athabasca tar sand of Canada (which you saw in Video Band 11) is an example of such a deposit.

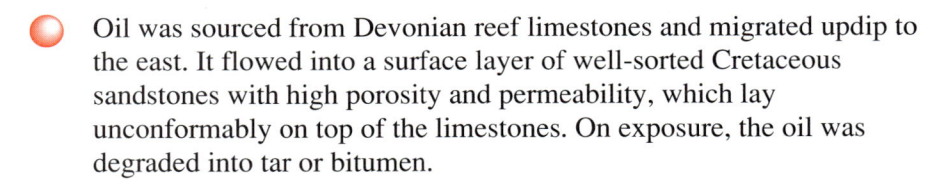

- How did the Athabasca oil sands form?

- Oil was sourced from Devonian reef limestones and migrated updip to the east. It flowed into a surface layer of well-sorted Cretaceous sandstones with high porosity and permeability, which lay unconformably on top of the limestones. On exposure, the oil was degraded into tar or bitumen.

The resource potential of oil sands is huge — the Athabasca deposit alone is estimated to contain as much oil as the conventional reserves in the whole of the Middle East.

Pitch lakes are deposits of bitumen formed from oil that has seeped to the surface, accumulated in large depressions and become solid. Examples include deposits in Trinidad, and the Bermudez Lake in eastern Venezuela which has estimated reserves of 6.1 million tonnes. When compared with the Brent oil and gas field (which initially contained about 290 million tonnes of oil), or with oil sands, this is a very modest resource indeed.

4.7.2 Orimulsion

Even underground, when oil is exposed to air or water for periods of time of the order of tens of millions of years, bacteria attack and degrade it to bitumen or tar. The heavy oils that result are extremely viscous and are not suitable for production by normal oil extraction methods, but nevertheless can represent a considerable hydrocarbons resource. Because they still lie deep underground they cannot be mined like the Athabasca oil sands or the oil shales of the Midland Valley of Scotland. To produce oil, heating has to be accomplished below ground, but the rock is not inflammable like the oil shales of Utah, so underground distillation by firing (as in the example in Video Band 10) is usually not possible.

The large bitumen reserves in the Orinoco region of Venezuela are being exploited by injecting water or steam downhole to produce an emulsion of bitumen and water. The resulting emulsion is known under its trade name of **orimulsion** (from *Ori*noco and e*mulsion*). Emulsifying the bitumen underground lowers its viscosity and enables it to be pumped to the surface. Since the emulsion is a liquid, it can be pumped through pipelines and stored or transported in tanks like conventional crude oil. The fuel properties of orimulsion are close to those of coal, and so it is usually quoted as a resource in tons of coal-equivalent.

4.7.3 Gas from the deep-sea bed

Sea-floor sediments often contain organic matter that will naturally decay to form hydrocarbons. Methane is constantly produced on the sea floor. Unless it is trapped in some way, this methane bubbles up to the surface and escapes into the atmosphere.

The temperature and pressure of the deep oceans worldwide are controlled respectively by cold polar currents that move along the ocean floor, and by the weight of water above the sea bed. The deep-sea bed is consequently at a temperature of around 1–2 °C and a pressure of several hundred times atmospheric pressure. Under these physical conditions, a frozen form of methane in water called **gas hydrate**, can form. Appropriate temperatures and pressures are found on the sea bed where water depths are greater than about 400 m, which in turn means that gas hydrates can form over most of the ocean floor.

Gas hydrates form freely in the oceans but they are not found as a carpet on the ocean floor. This is because gas hydrates are lighter than sea water (they have a density of 850 kg m^{-3}). Normally as soon as they form, they float upwards, turning back into methane and water in the lower pressures and warmer temperatures of the upper layers of the ocean.

However, within the sediments just beneath the ocean floor, crystals of gas hydrates form and are buoyed upwards. Frequently the rising crystals form a 'log-jam' within the pore spaces of the sediment. Once this occurs, more gas hydrate crystals will become trapped in the pore spaces of the sediment underneath.

Gas hydrate crystals also readily absorb methane into their lattice. Fully saturated gas hydrates can hold up to 200 times their own volume of methane within their crystal structure. Eventually, the spaces between grains of sediment become completely filled by gas hydrate crystals which themselves are saturated in methane. Saturated gas hydrate crystals are denser than seawater, so the zone as a whole is gravitationally stable.

The best conditions for forming gas hydrates are to be found in areas of the ocean where relatively thick carpets of sediment are being deposited and where large quantities of methane are also being produced. The Atlantic

(a)

(b)

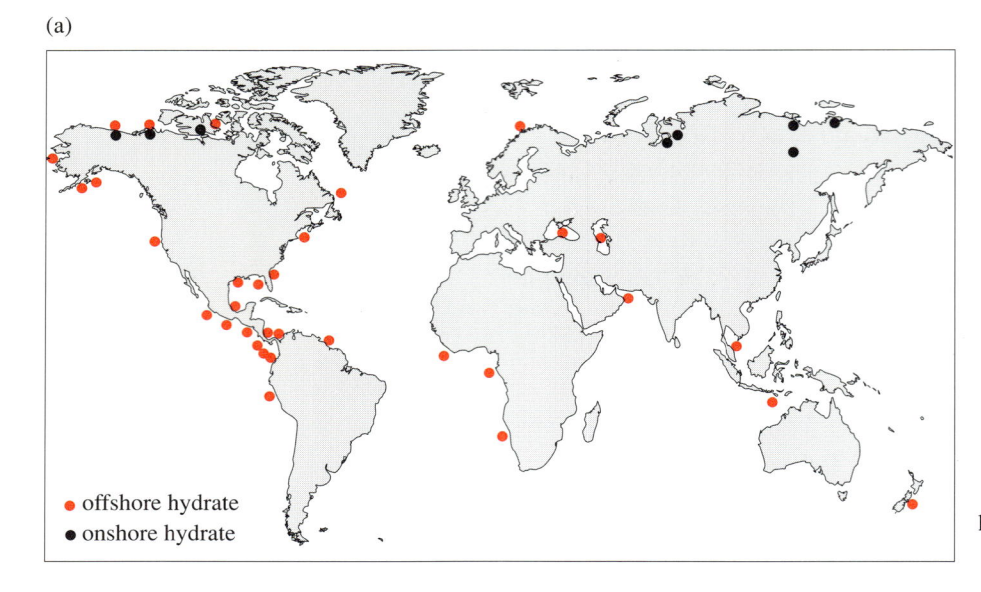

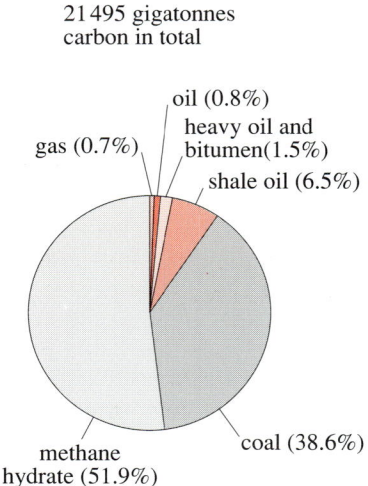

21 495 gigatonnes carbon in total

- oil (0.8%)
- heavy oil and bitumen (1.5%)
- gas (0.7%)
- shale oil (6.5%)
- coal (38.6%)
- methane hydrate (51.9%)

• offshore hydrate
• onshore hydrate

margins of the Americas, Africa and Europe have relatively high rates of sedimentation, plenty of buried organic matter and adequate heat flows and gas hydrates have been recognized from the Arctic, the North Atlantic, the north Pacific Ocean near the Aleutian Islands and in the Bering Sea (Figure 64a).

The deep-sea bed is not the only source of gas hydrates. Permanently frozen ground which gives permafrost conditions can also be suitable for hydrate formation. The Arctic zones of Russia and North America and the unexplored continent of Antarctica are prospective areas (Figure 64a).

Current estimates suggest that gas hydrates might contain some 10^{16} tonnes of carbon (Figure 64b), or slightly more than the amount of carbon that is locked into all the other fossil fuels added together. However, recovery of gas hydrates from both the deep-sea bed and permafrost zones is both expensive and technologically difficult. Despite these vast resources, using gas hydrates as a fuel lies a long way into the future.

Figure 64 (a) Known global occurrences of methane hydrate. (b) Percentage of the total carbon in methane hydrate and other fossil fuels.

4.8 Oil and gas reserves worldwide

Just as some countries have no coal while others have vast reserves, so it is with petroleum. Again, there are sound geological reasons for this. The most significant is a lack of sedimentary basins within the jurisdiction of the country. Source rocks are confined to *continental crust* (Block 1, Section 3). A few countries (Iceland is perhaps the best example) are made entirely from oceanic crust. They have no source rocks and therefore no petroleum. Similarly, igneous and most metamorphic rocks cannot source and rarely host petroleum, so countries made almost entirely from these rocks, like Sweden, are poor in petroleum resources.

Countries rich in petroleum resources generally have one of the following two features.

(a) They have large continental or continental shelf areas, which are statistically more likely to contain one or more sedimentary basins, e.g. USA (3.1% of 1993 world oil reserves), Russia (4.8%) and China (2.4%).

(b) For geological reasons they happen to include prolific, hydrocarbons-rich basins within their borders. The 'top seven' countries include Saudi Arabia (25.9% of 1993 world oil reserves), Iraq (9.9%), Kuwait (9.6%), Iran (9.2%), Abu Dhabi (9.1%), Venezuela (6.3%) and Mexico (5.0%).

4.8.1 Estimating reserves

There are inherent difficulties in estimating petroleum reserves accurately. In the case of coal, economic seams can be proved by drilling, and estimates of the volume of coal reserves can then be calculated by multiplying the thickness of coal in workable seams by the area over which they have been identified. This is not possible with petroleum. The volume of oil and gas is never as simple as the volume of mature source rock, or the potential reservoir volume, or known trap volume. It is a subtle combination of all these. Estimating reserves of oil and gas is therefore a more complicated and less precise process than estimating reserves of coal.

How then are estimates made of how much oil and/or gas any given sedimentary basin will contain? In answering this question, we come to understand just how important the play concept is to petroleum reserve analysts. Take the southern North Sea as an example. During the Second World War, nothing was known of the Permian gas play in that area (you can be certain that the course of that war would have been different if it had). Analysts of the time would have said that there were no proven reserves of gas in the southern North Sea.

When the Groningen field was discovered in 1959, and with it the new play, analysts said that the gas reserves of the area were some 3×10^{12} m^3, with the prospect of much more gas being discovered in the new play in other parts of the region. Twenty-five years and over fifty discoveries later, gas reserves in the region were estimated to be over 5.5×10^{12} m^3, almost twice the 1959 figure. This illustrates a common pattern in many basins; a rapid rise in reserves estimates when the play is first discovered, followed by a steady decline in the rate of increase as the play is explored.

Figure 65 shows a typical pattern of reserve assessment over time in a basin in which there are several petroleum plays. Typically, large fields are discovered early in the life of a play because they lie in the biggest structures, and these are the ones that are identified and drilled first (Figure 65, point A). Reserve estimates increase rapidly. Eventually, at B, only smaller fields within the play remain undiscovered and the rate of reserve increase declines. If only one play is ever discovered in this basin, then the total reserves discovered over time will approach the amount B′. However, if a second play is discovered (C), reserve estimates increase rapidly again, spurred on now by the new discoveries of both plays added together. The total reserves for the

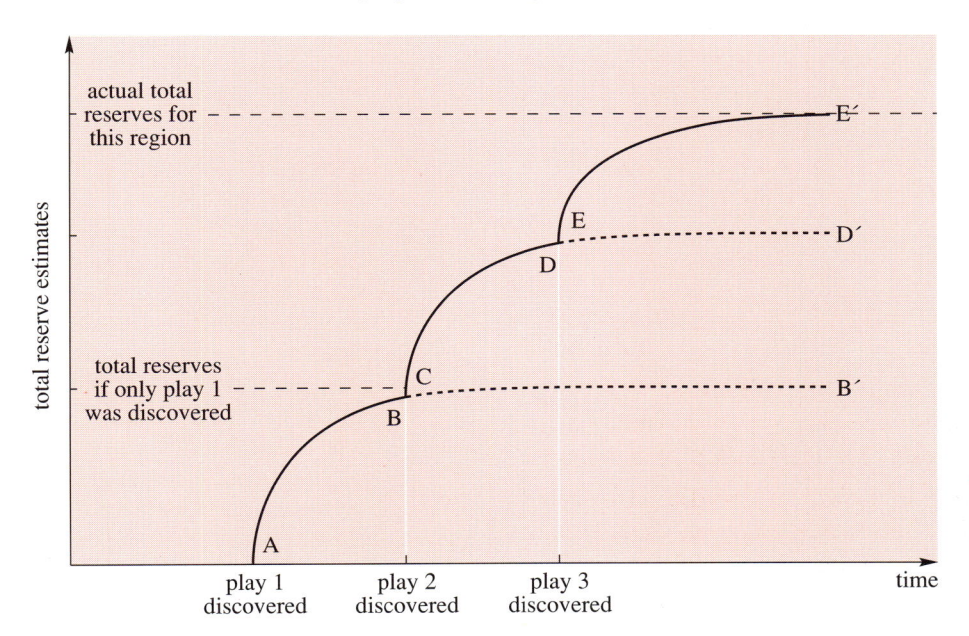

Figure 65 The increase in reserves when new plays are discovered and evaluated in any sedimentary basin. See text for discussion of letters.

basin as a whole will now approach D′ with time. Discovery of a third significant play, E, will increase the reserve estimates again, and now the basin will have total reserves estimated at value E′.

Reserve estimates for any area are therefore sensitive to how exploration is conducted. An accessible area with no technological barriers to exploration and many participants (with good ideas) will yield its true reserve potential quite quickly, perhaps in ten or fifteen years. On the other hand, sedimentary basins located in areas of difficult access (geographic or political) or with technological barriers to be overcome will continue to surprise reserves analysts.

4.8.2 UK petroleum reserves

Figure 66 shows the location of the main sedimentary basins and therefore the principal oil and gas plays in the UK area. Most of the UK's oil and gas reserves lie under the North Sea. Gas has also been discovered under Morecambe Bay in north-west England and both oil and gas in the area west of the Orkney Islands. Onshore, small but significant oilfields have been discovered near Eakring in Nottinghamshire, in the Wessex basins of southern England and near Edinburgh in Scotland. Outside these areas, most of onshore UK does not contain any important petroleum prospects.

The major British plays are numbered in Figure 66, and details of play components are given in Figure 67.

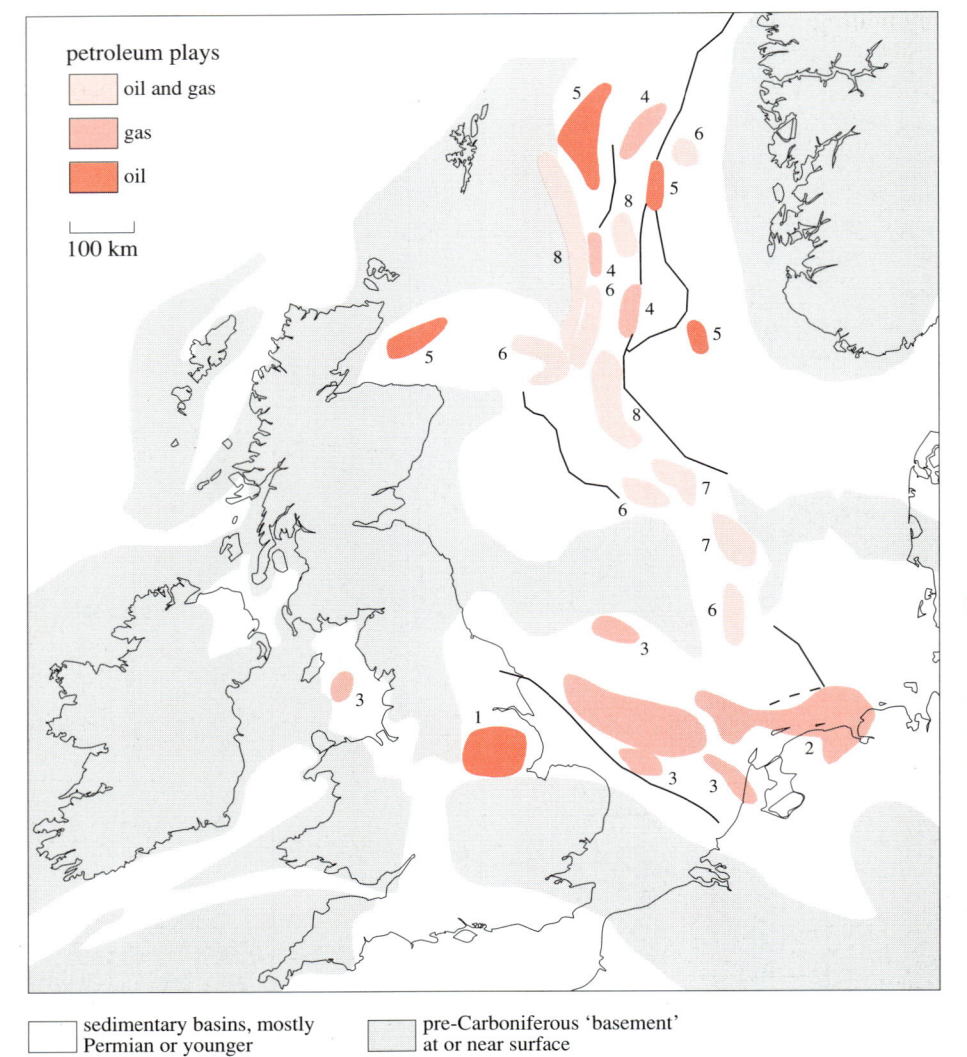

Figure 66 The geographical distribution of the main sedimentary basins and petroleum plays in the UK and its continental shelf. The grey area indicates that rocks unsuitable for petroleum generation lie at or close to the surface; these areas include much of Scotland, Wales and Ireland, as well as Norway and Belgium. The numbered and coloured areas are the important oil and gas plays which are identified in Figure 67.

Three major groups of source rock are important:

1. the Carboniferous coals, which source gas (particularly in the southern North Sea and Morecambe Bay) and shales which can source oil;
2. the Jurassic Brent Group coals and shales, which source some gas and oil in the northern North Sea;
3. the Jurassic Kimmeridgian shale, which is the most important oil source rock in north-west Europe but which also sources gas in some northern North Sea fields.

Reservoir rocks occur at many horizons. Carboniferous sandstones are oil and gas bearing onshore (play 1 in Figure 67). More significant are the Permian and Triassic desert sandstones of Groningen and the southern North Sea (plays 2 and 3 in Figure 67). Triassic–Jurassic marine sandstones contain large amounts of oil and gas in the northern and central North Sea areas, including the giant Brent and Statfjord fields (plays 4–6). The Cretaceous Chalk hosts petroleum in the central North Sea, notably in the Ekofisk field (play 7). Tertiary deep-water sandstones yield oil and gas in the Frigg and Forties fields of the northern North Sea (play 8).

Shales are common in the Tertiary and Lower Cretaceous of the central and northern North Sea area, and form good seals. Permian and Triassic salts are important seals in the southern North Sea.

Most of the important traps are related to the tectonic history of the North Sea area. Crustal stretching during the Jurassic–Cretaceous period produced a series of rifts through the central and northern North Sea. Many of these traps, like Brent, are generated by faulting which took place at this time. In the southern North Sea, dome structures associated with flow of Permian and Triassic salt form the major traps.

At the end of 1993, oil and gas fields in the UK, primarily in the eight plays shown on Figure 67, were estimated to contain in total 0.6×10^9 tonnes of crude oil and 0.6×10^{12} m^3 of natural gas. In the UK in 1993, R/P ratios were 6.1 for crude oil and 9.7 for natural gas.

Figure 67 The major components of UK's eight most important oil and gas plays, showing the age and nature of important source, reservoir and seal rocks. Important fields within each numbered play are: 1, Eakring, Nottingham; 2, Groningen and West Sole; 3, Morecambe Bay; 4, Statfjord; 5, Brent; 6, Piper; 7, Ekofisk; 8, Forties and Frigg.

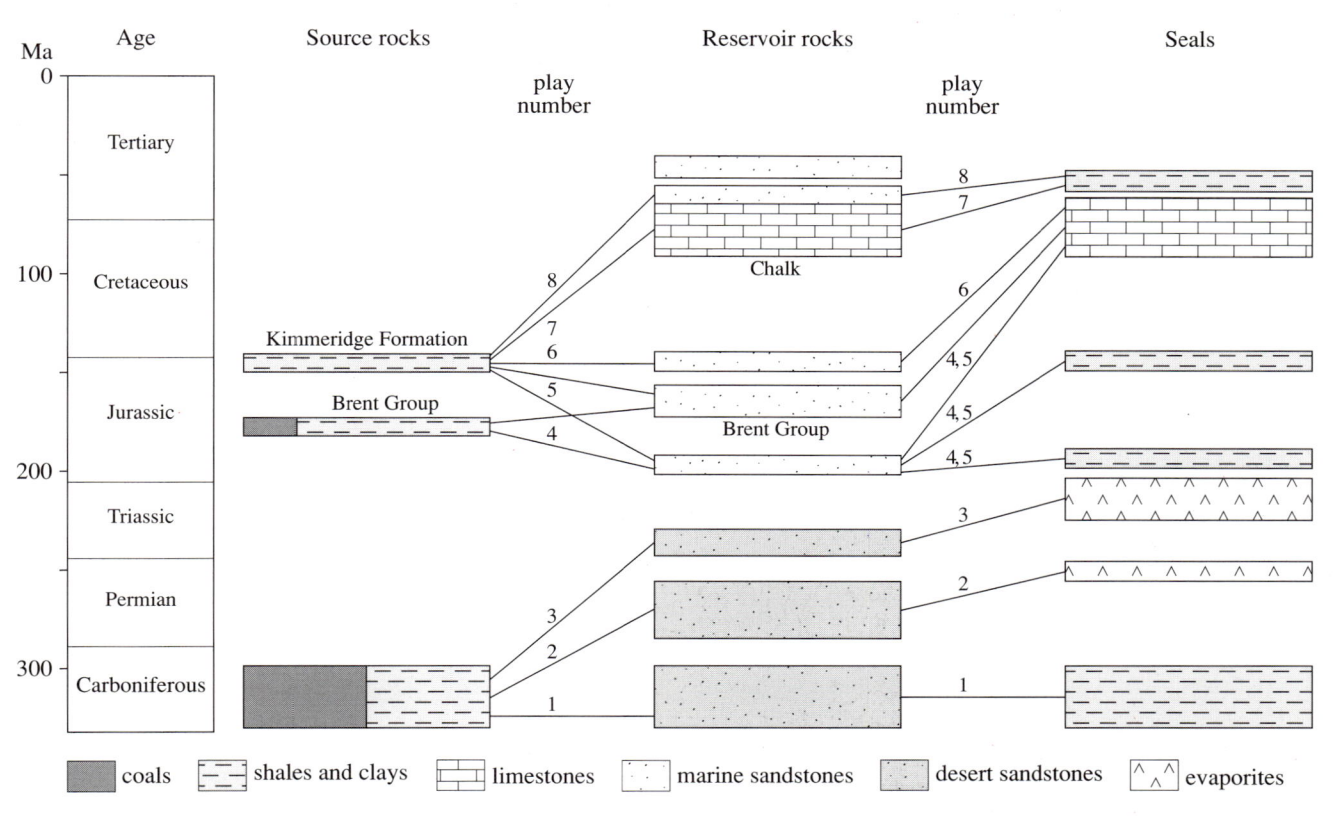

○ Does that mean that Britain will have run out of oil by the year 2000?

● No. Reserves lifetimes can't be used in this way to make such sweeping statements. Some fields, like Brent, will continue to produce oil and gas for ten to twenty years beyond that date. It does however suggest that British oil and gas production can be expected to decline in the early part of the twenty-first century.

4.8.3 World oil

At the end of 1993, world proven crude oil reserves were estimated at 136.7×10^9 tonnes. Figure 68 shows the breakdown of this world reserves total by continents. The Middle East had 66% of world oil reserves.

World R/P ratios rose sharply in the late 1980s, because dramatic increases in new reserves discovered in the Middle East outpaced the steady growth in world production. The greatest proportional demand for energy from oil at present comes jointly from Latin America and the Middle East: oil supplies about 65% of energy needs in both regions. In Eastern Europe, oil provides only 26% of energy, whereas in Western Europe it provides 46%.

The Middle East is the largest net exporter of oil, accounting for 53% of world crude oil exports. Western Europe is the largest importing region, accounting for almost a third of world crude oil imports.

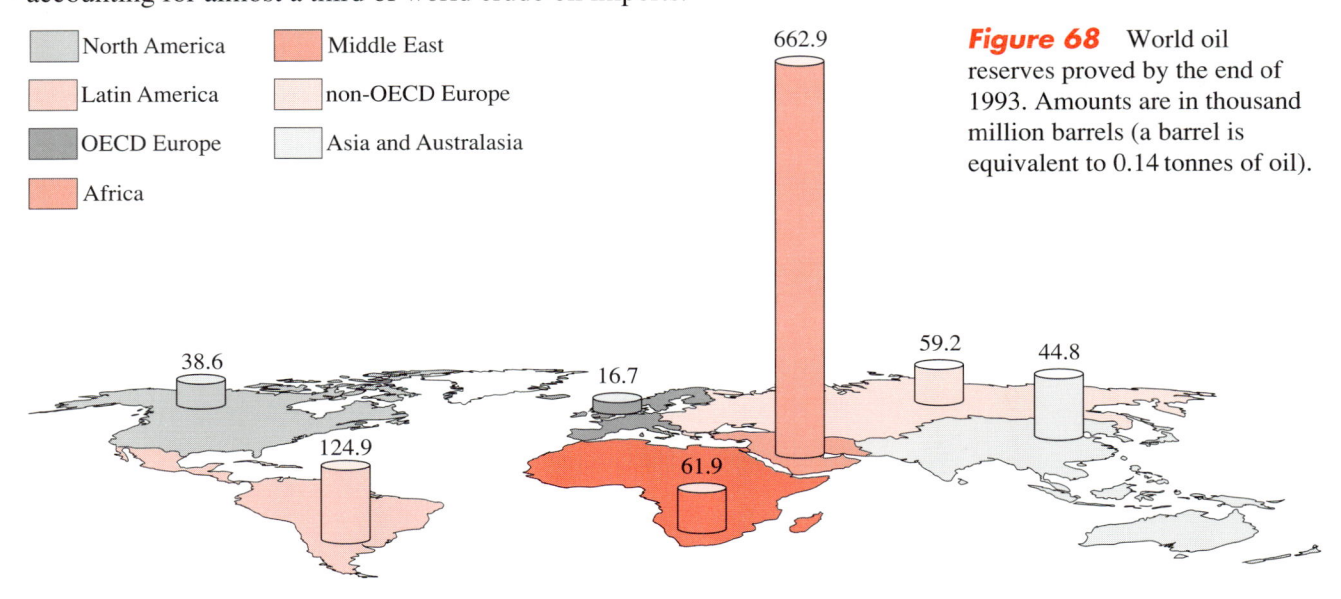

Figure 68 World oil reserves proved by the end of 1993. Amounts are in thousand million barrels (a barrel is equivalent to 0.14 tonnes of oil).

Oil in the Middle East

Two-thirds of the world's oil is to be found in the Middle East basin. Half the world's total has been found in just four countries; Saudi Arabia, Kuwait, Iran and Iraq. By 1993, Saudi Arabia alone had identified reserves equivalent to one-quarter of the world's total, enough to last 84 years at the 1993 rate of production. Why is the Middle East such a prolific oil province?

From 250 to about 30 Ma ago, the African and Eurasian continental plates were separated by a large ocean, called *Tethys*. The Middle East basin was part of a broad continental shelf up to 2000 km wide which bordered the southern side of this ocean. During the whole of this time the region lay in tropical latitudes. Organisms were abundant in the shallow continental shelf seas.

Shales rich in organic matter that were to become good source rocks accumulated in localized depressions on the continental shelf. On other parts of the shelf, shallow-water coarse carbonate sands were deposited, having a natural high porosity. At times, sea level stood low, evaporation took place in the hot climate, and widespread salt deposits developed which capped the potential carbonate reservoirs. In all, a thick sequence of carbonate and evaporite sediments was deposited on the shelf.

When the African and Eurasian plates collided together about 30 Ma ago, Tethys ceased to be an ocean and the continental shelf sediments that once surrounded it were folded and uplifted. This uplift resulted in the formation of the Zagros Mountains of Iran (Figure 69), the formation of anticlines suitable for trapping oil in the shelf sequence deposited earlier, and tectonic burial of the source rocks.

In the Middle East basin petroleum traps were formed in three ways:

1. Faults in the Precambrian basement resulted in the large upstanding fault blocks (Figure 69) over which sediments were draped to form broad anticlines.
2. The early Cambrian salt deposits moved to form salt plugs, resulting in doming of overlying sediments.
3. Anticlinal folds formed, and petroleum generated from Mesozoic sediments migrated upwards into the anticlinal traps.

A prominent north–south trend in the fields can be related to the direction of the faults in the Precambrian basement. This faulting gave rise to tilted fault blocks which were elongated in a north–south direction. So any anticlinal traps formed by sediments draped over these uplifted fault blocks would also be elongated north–south. The rounded outlines of fields in the Persian Gulf are related to salt plug structures, whereas the north-west to south-east trend of fields in the Zagros Mountains is related to similarly oriented anticlinal folds in the late Tertiary fold belt.

One of the largest oilfields was discovered in Kuwait in 1938 and is known as the Burgan field (Figure 69). The trap is an elongated dome and its potential was realized because the crest of the anticline exposed at the surface was marked by bitumen seepages.

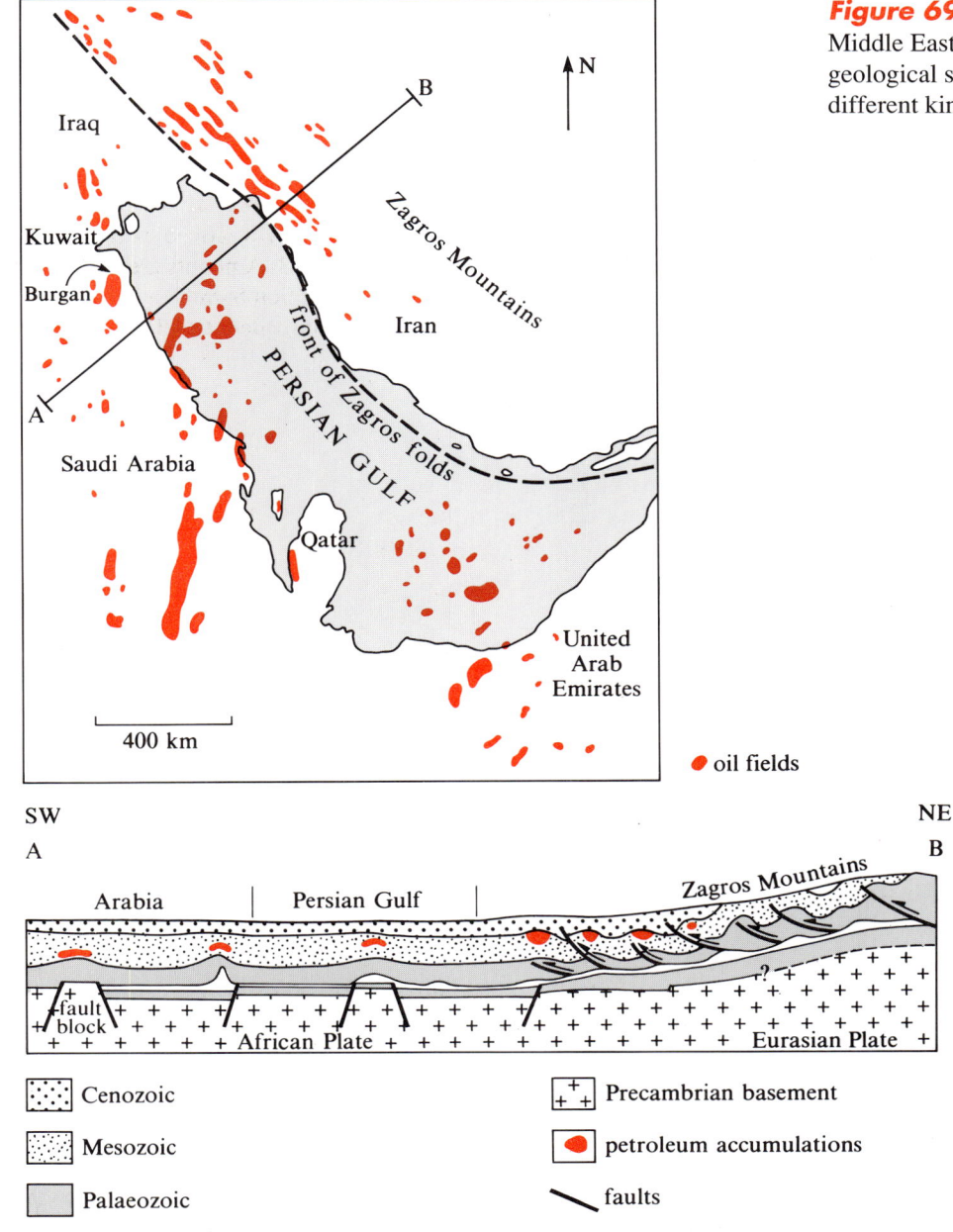

Figure 69 The oilfields of the Middle East basin, and a SW–NE geological section to illustrate the different kinds of trap.

4.8.4 World gas

At the end of 1993, world proven natural gas reserves were estimated at $142.0 \times 10^{12} \, m^3$. Figure 70 shows the breakdown of this total by continents. The Commonwealth of Independent States (CIS) and the Middle East together held 70% of the world's natural gas reserves.

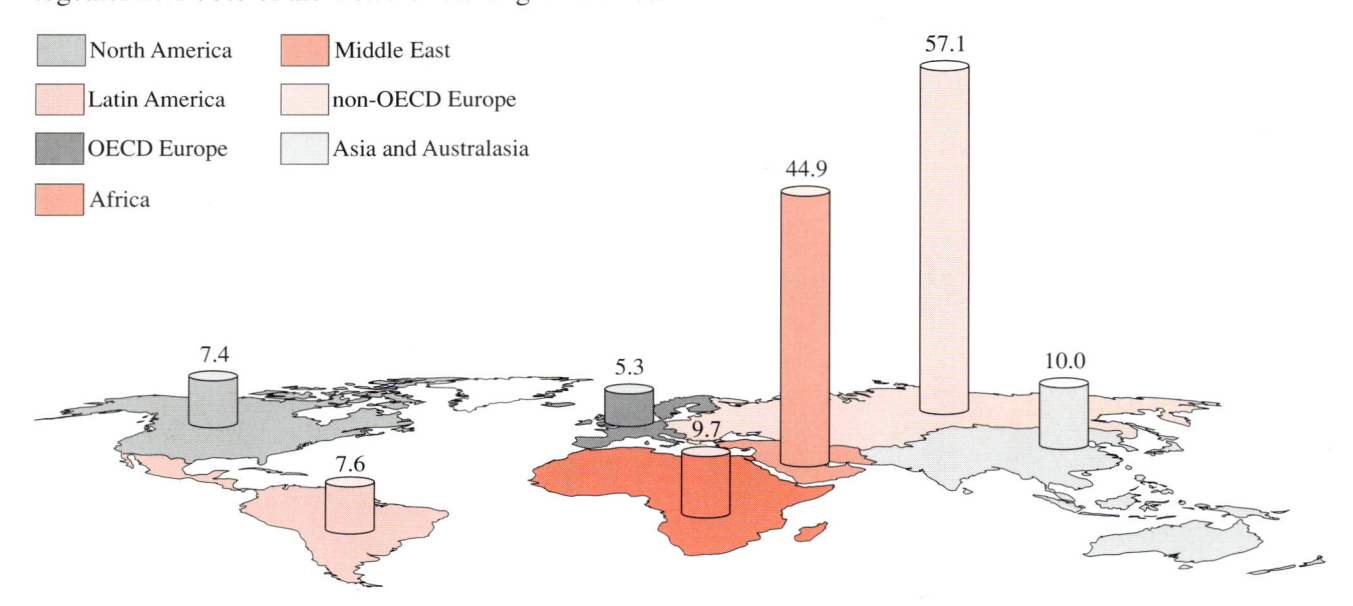

- North America
- Latin America
- OECD Europe
- Africa
- Middle East
- non-OECD Europe
- Asia and Australasia

Values shown: 7.4, 7.6, 5.3, 9.7, 44.9, 57.1, 10.0

Figure 70 World gas reserves proved by the end of 1993. Amounts are in trillion (10^{12}) cubic metres.

The early 1990s saw steady increases in gas reserves in both the CIS and the Middle East. In Asia and Australasia, gas provides only 7% of energy, whereas in Western Europe it provides 16%. The greatest demand for energy from gas at present comes from Eastern Europe, where gas supplies about 40% of energy needs.

World gas reserves increased steadily from 1966 to 1993, the R/P ratio rising over that period from 39 to 65. World gas production also continued to rise despite a decline in output from USA and Russia, the two largest producers. In fact, in the early 1990s the gas industry worldwide was finding four times as much gas as was then being consumed. At present, the largest net movements of gas are from Russia to Western Europe, and from Canada to the USA.

4.8.5 Reserve predictions from discovery and production curves

So far we have relied on the play as a basis for describing and estimating reserves. By way of contrast, it can be useful to look at reserves in terms of discovery *rates* rather than at total amounts. How can this type of data be used to predict reserves? In 1956, Dr M. King Hubbert produced mathematical models describing the life cycle of oilfields. The technique is described by Hubbert himself in Video Band 10, and pictorially in Figure 71.

Now rewatch the first part of Video Band 10: *Oil — Finds for the Future* (the first 11 minutes or so).

Figure 71 shows schematically three curves relating US oil discovery and production over time. The first shows changes in the amount of oil that has been discovered in America from 1900 to the present time and gives a prediction for the early part of the twenty-first century (the discovery curve). The second shows the amount of new oil produced (or expected to be produced) each year over the same time period (the production curve). A third shows the rate at which known reserves are changing. Figure 71 shows that oil discoveries began in the USA in the second half of the nineteenth century, and reached a peak in the 1950s. Production naturally started later than discovery, and also reached its peak later, in the 1970s. In the early years of US oil industry, discovery exceeded production and reserves increased. As time progressed, fewer new plays were discovered and production began to outstrip discoveries. At that time, known reserves began to decrease.

- From Figure 71, did US oil production exceed discovery in the 1940s or the 1960s?

- The 1960s, indicated by the crossover between the discovery and production curves. That time also marks the change from *reserves increasing* to *reserves decreasing*, the shift from positive to negative in the 'change in proved reserves' curve.

- Then what does the peak (1940s) and the trough (1980s) in the 'change of proved reserves curve' indicate?

- The 1940s peak indicates a maximum in the change of reserves — reserves were increasing fastest then. The 1980s trough indicates a minimum in the change of reserves, when reserves were decreasing fastest.

The mid-point of the life of a petroleum province is therefore given by the time when reserves stop increasing and start decreasing, in other words the time when the 'change of reserves' curve in Figure 71 passes from the positive to the negative part of the graph.

Production and discovery rates can be plotted against time, just as in Figure 71, for any field or petroleum province. The rate-of-change curve can then be used to define the reserve mid-point, which in turn will give the expected life of the reserves in that field or province. The area under the discovery curve gives the total petroleum reserve for that field or province. Reserve estimation using this type of analysis is called the *Hubbert technique* after its inventor. The Hubbert technique can be used for small groups of fields through to global oil reserves.

In Video Band 10, Dr King Hubbert uses his own technique to show that US reserves lie between a pessimistic value of 150 billion barrels of oil and an optimistic one of 200 billion barrels. The US oil industry passed its discovery peak in 1970, so it can be predicted to be viable until between the years 2040 and 2060. If the Hubbert technique is used to determine global oil reserves it shows a discovery peak that was passed in 1980.

How do the Hubbert curves relate to proven reserves, the R/P ratio and the reserves lifetime? Proved reserves can be calculated from the Hubbert discovery and production curves for any particular time; reserves are given simply by discoveries (i.e. the area under the discovery curve, up to that time)

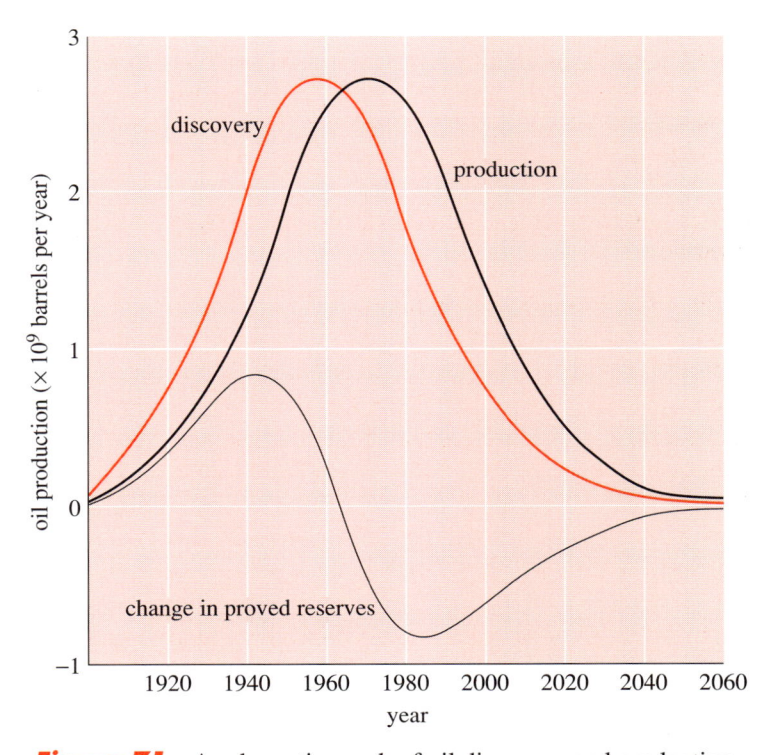

Figure 71 A schematic graph of oil discovery and production amounts against time, illustrating the Hubbert model for the discovery, production and rate of change of proven reserves of oil in the USA.

minus production (the area under the production curve, up to that time). The R/P ratio is the total amount of reserves *divided* by the amount of production at that time (a *point* on the production curve). This value is also the reserves lifetime under those fixed production conditions.

The Hubbert 'change in proved reserves' curve in Figure 71 is subtly different from either the R/P ratio or the reserves lifetime. The 'change' curve at any time is the rate at which discoveries are being made (a point on the discovery curve) *minus* the rate at which oil is being produced (a point on the production curve). This tells us whether the total reserves are increasing or decreasing, and how the rate of increase or decrease compares with previous years. The 'change' curve therefore shows the 'dynamic health' of reserves (whether they are increasing or decreasing) whereas the R/P ratio or reserves lifetime shows the 'static health' of reserves (how long they would last at a fixed rate of production).

Try Activity 5 to investigate the 'static health' of UK oil reserves.

Activity 5

Figure 72 shows Hubbert curves for UK's oil discovery and production from 1969 to 1992.

(a) What, on Figure 72, does the area of one of the squares represent? How could we use these squares to give the total amount of oil discovered in UK between 1968 and 1992? When did UK oil reserves stop increasing?

(b) Estimate the amount of oil discovered and the amount of oil produced in the UK from 1968 to 1992. Estimate the amount in tonnes of UK oil reserves in 1992 (i.e. by how many tonnes UK oil discoveries by 1992 had exceeded total production by 1992).

(c) By the end of 1992, *proven* oil reserves in the UK stood at 4.1 billion barrels (a barrel is equivalent to 0.14 tonnes of oil). Figures published by UK Department of Energy in 1989 estimate that there are 3.8–13 billion barrels of recoverable oil as yet undiscovered in the UK. Production in the UK sector in 1992 averaged 1.89 million barrels of oil per day. Work out the UK's 1992 R/P ratio for *proven reserves*, and the best- and worst-case *expected* reserves lifetimes for UK oil, based on the Department's 1989 projections and 1992 production rates.

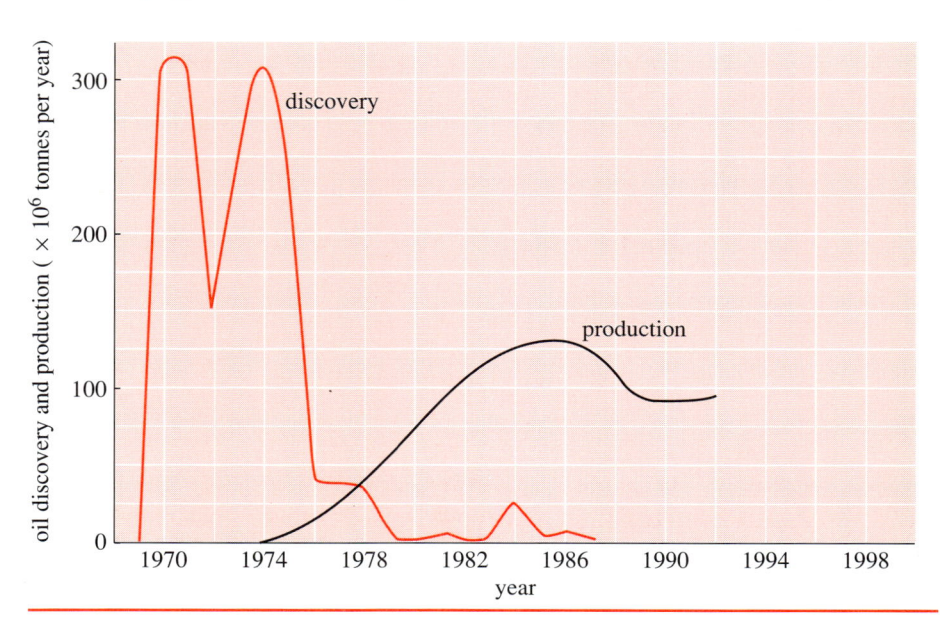

Figure 72 Hubbert curves for UK's oil discovery and production from 1969 to 1992. The UK has already passed its mid-point as far as known oil plays are concerned.

Reserves lifetimes calculated in the early 1990s therefore range between 5.9 years (actual 1992 figures) and about 11–25 years (worst and best case predictions). These are 'static' lifetimes, calculated for a particular year. But reserves evolve in a dynamic way; for example depletion of reserves often leads to a price rise, stimulating and funding research and exploration, which in turn leads (or has always led so far) to an increase in reserves. Whether you accept these dates as indicating an end to UK oil depends largely on your optimism about (a) recovering more oil from existing UK fields, (b) finding new plays in the established UK basins, or (c) finding recoverable plays in the Atlantic margin — the UK's **frontier area** (unexplored area believed to contain fossil fuels).

One thing remains clear: oil and gas are finite resources. Wholesale discovery of new plays will extend the lifetime of oil but not indefinitely, as Dr King Hubbert was able to show. We must be drawing ever closer to the end of the 'hydrocarbons age'.

4.9 Summary of Section 4

1. Fluid petroleum migrates for two reasons. As a sedimentary sequence is buried, the sediments compact, losing pore space between grains. Water and petroleum are thus expelled from between the mineral grains. Excess pressure also helps to expel hydrocarbons from the source rock as it matures. Petroleum moves in response to the pressure gradient set up within the rock. Fluids can move upwards, sideways or even downwards, depending on the detailed local pressure gradient.

2. For petroleum to become concentrated enough to represent a substantial oil or gas accumulation, it must be retained within suitable reservoir rocks. Reservoir rocks must be porous enough to contain substantial amounts of petroleum and they must also be permeable enough to allow fluids to flow into and out of the reservoir. Igneous and metamorphic rocks are usually unsuitable reservoir rocks; almost all reservoir rocks are sedimentary rocks.

3. Permeable reservoirs must be capped by impermeable seals, usually shales and evaporites.

4. Structural traps are formed by the deformation of sedimentary rocks and include anticlinal domes and fault traps. Other examples of structural traps occur, including traps associated with salt plugs. Stratigraphic traps occur where there is a permeability barrier caused by variation in sedimentary rock types.

5. Combination traps are often formed when folded or faulted reservoirs are overlain unconformably by shales, which act as a seal. The Brent field is an excellent example. About 78% of the world's crude oil resources are held in structural traps, 13% in stratigraphic traps and 9% in combination traps.

6. Geophysical techniques that allow whole regions to be surveyed are relatively inexpensive compared with drilling and detailed surveys, but locate only broad areas which might be suitable for exploration. Gravity surveys can be used to locate sedimentary basins because substantial volumes of sediments are characterized by lows in the Earth's gravitational field. Sedimentary basins are also characteristically areas of uniform geomagnetism, whereas igneous rocks and old metamorphic basement rocks often have a highly variable magnetic structure.

7. The main method of determining whether an area has traps that may contain petroleum is through seismic surveying. Seismic sections

provide a continuous survey of sedimentary sequences, major fault locations, and other structural information. Once detected, a potential trap can be mapped in detail using 3-D seismic surveys and evaluated by drilling. Once the trap shape, the thickness of the oil- and gas-bearing parts of the reservoir, the porosity and permeability of the reservoir rock have been determined, the volume of oil and gas that can be recovered from the field can be estimated.

8 Primary recovery methods produce at best only 30% of the reserve. This can be boosted to 40–50% by secondary recovery techniques of pumping water and gas under pressure into the reservoir. Tertiary or enhanced recovery is intended to increase recovery still further, mainly by the injection of detergents into the reservoir to reduce viscosity. Another exciting prospect is the use of bacteria to produce gas or solvents within the reservoir. Any improvement in the percentage recovery will have an important bearing on the estimates of total recoverable reserves.

9 The most significant local environmental risks from petroleum production come from accidental spillages and transportation.

10 Petroleum plays hold the key to resource evaluation. Once a new play has been discovered, the elements of that play can be assessed and the likely petroleum reserves can be calculated. If our current petroleum reserve calculations turn out to be inaccurate, it is most likely to be because significant plays have not yet been discovered.

11 The oil and gas fields in the North Sea are mainly held in Permian (gas), and Jurassic and early Tertiary (oil) reservoirs. The source of the gas in the southern North Sea basin is the Carboniferous coals, and the Jurassic Kimmeridge shales are the source rocks for oil and gas in the northern North Sea basin. Some small gas and oilfields have been discovered onshore in Britain.

12 The UK reserves in the North Sea are minute compared with those of the Middle East which has 66% of the world proven reserves of crude oil. Ideal conditions prevailed in the Middle East during the Mesozoic–Cenozoic for the deposition of enormous volumes of source rocks which underwent maturation after the formation of numerous large traps.

13 In 1993, the world had reserves of 137 billion tonnes of oil which at 1993 rates of production would last until the year 2036, and 142 trillion cubic metres of deep gas which at 1993 rates of production would last until the year 2058.

14 Large amounts of oil and gas are locked into oil sands and gas hydrates, but they do not yet constitute an economic resource.

Question 28

In Figure 73, at which of the points 1 to 7 would you expect petroleum to accumulate in the reservoir? (Assume that the faulting has not hindered migration of oil and has not permitted oil to escape to the surface along the faults.) Where would you look for an oil seepage?

Figure 73 A hypothetical cross-section through some typical trap locations.

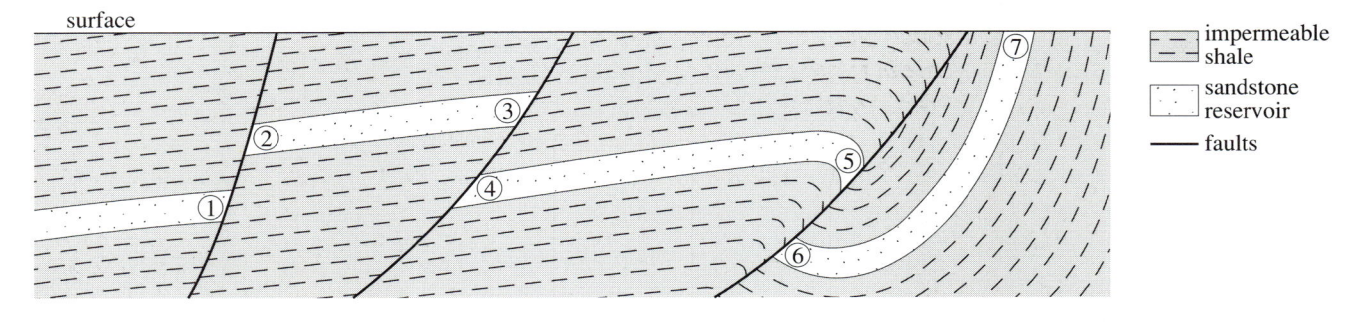

Question 29

List the techniques that could be used to find an offshore oil trap and to test if there is oil within it, and the methods used to determine the shape of the trap and reservoir.

Question 30

What data would you require in order to produce an estimate of reserves of oil remaining in an oilfield that had been in production for 5 years?

Question 31

Study Figure 74 and answer the following questions.

(a) An oil well has penetrated the oil-bearing sandstone A, and with falling production from the field it has been decided to use fluid injection to improve recovery. Where would you site four boreholes to achieve this? (Two would be used to obtain water and gas, and the other two to pump water and gas down into reservoir A.)

(b) What kinds of trap are shown in Figure 74?

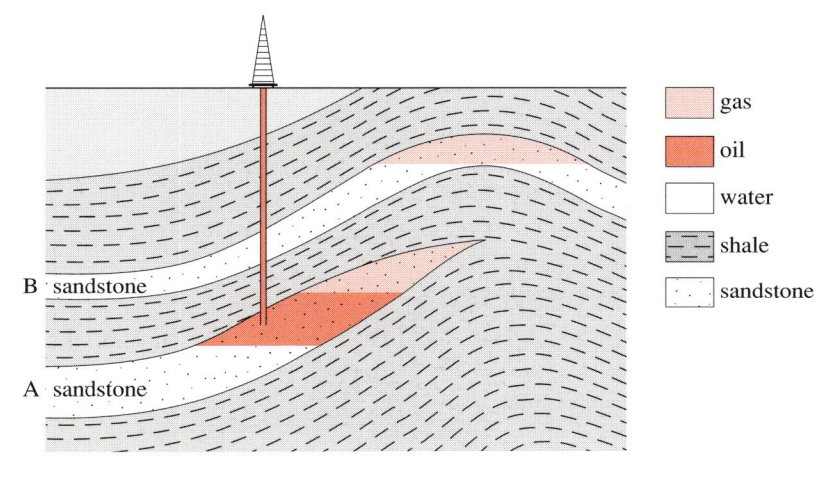

Figure 74 An oil well penetrating two sandstone reservoirs.

Question 32

Why has the Middle East got such enormous conventional petroleum reserves?

5 THE FUTURE FOR FOSSIL FUELS

Can we continue to use fossil fuels in the way that we do presently? This is the fundamental energy question that faces us in the late twentieth century, and it is perhaps best considered in two parts. First, what is the lifetime of fossil fuels, given our present rates of discovery, production and methods of use? Second, what are the implications of burning fossil fuels which may limit or restrict their use, or promote a change of energy policy? We shall address each of these questions in turn.

5.1 World energy use and fossil fuel lifetimes

What is the lifetime of fossil fuels? At first sight the answer seems straightforward, and is given by calculating the global R/P ratios for any fuel. In reality, the situation is not so straightforward. The following questions and answers should help to highlight some of the problems:

● If the current R/P ratio for any given fuel was a *reliable* predictor of the lifetime of that fuel, how might we expect it to vary over the next few years?

○ The R/P ratio should fall steadily over successive years, reducing by one each year.

Is that what happens for real commodities? The R/P ratios for global natural gas and crude oil since 1966 are shown in Figure 75.

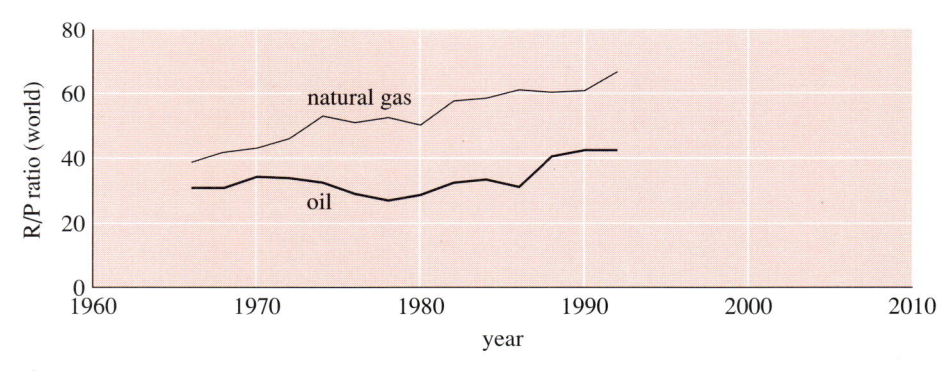

Figure 75 Global R/P ratios for crude oil and natural gas for the period since 1966.

● Did global oil and gas R/P ratios fall steadily from 1970 to 1990?

○ No. Both rose, gas more than oil.

Since 1966, gas and oil R/P ratios have been rising at an average rate of one per year and 0.2 per year respectively. To have predicted an end to world gas in the year 2006 based on a 40-year 'lifetime' in 1966 would have been misleading and unhelpful. It is similarly unreasonable to predict an end to world gas reserves in 2050 based on the 1990 R/P ratio shown in Figure 75.

Like reserves lifetimes, R/P ratios do not vary in a simple way year by year because production rates and proved reserves are not static quantities. R/P ratios reflect the complex interactions between amount of reserves, discoveries and discovery incentives, production, and supply and demand.

- For example, what will the discovery of a new economically viable oil or gas play do to (a) the current R/P figure, and (b) the R/P figure in succeeding years?

- Discovery of a new play immediately increases the reserves without affecting current production. (a) The current R/P ratio will increase sharply. (b) Over succeeding years, more discoveries within the play should add to reserves, continuing the rise in the R/P figure, but as production becomes established, the rise in R/P ratio will be checked and will begin to fall when production outstrips discovery.

Interpreting R/P ratios over time is therefore far from straightforward. If there is a large stock of reserves (i.e. large R/P ratios or reserves lifetimes) for any commodity, exploration incentives are low. Demand may increase, since the commodity is felt to be 'secure', and this may lead directly to an increase in productivity. High R/P ratios tend to fall over time. Conversely, low commodity stocks promote higher prices, which encourages exploration, leading to discoveries and increased reserves. Production may fall at the same time if suppliers wish to conserve stocks. Low R/P ratios therefore tend to rise over time. The level towards which R/P ratios are attracted is a complex interaction between commodity value, production lead times and the current political and economic situation worldwide.

- What effect did closing most of the UK's deep coal mines in the early 1990s have on UK coal R/P ratios?

- Most UK coal had already been discovered, so steady coal production would lead naturally to dwindling reserves and falling R/P ratios over time. If, however, coal production were to be drastically reduced, the R/P ratio should rise sharply. At the same time some coal that formerly was considered to be economic to produce was no longer economic. Some reserves became reclassified as resources, and total reserves decreased, leading to a fall in R/P ratio. Intuitively, a severe loss of production should mean that the R/P ratio should rise overall, but it is far from clear whether 'lifetimes' increase, decrease or even remain constant.

The point being made is this: current R/P figures and reserves lifetimes indicate the life of any commodity only when unrealistic assumptions are made, such as no future discovery or constant future production levels. Rather, R/P ratios indicate the 'state of health' of a commodity and more usefully predict whether the industry surrounding that commodity is likely or not to be spurred into exploration.

- What circumstances, then, might bring about an end to the life of any commodity such as coal, oil or gas?

- Two sets of circumstances might: the limit of exploration where no more reserves remain undiscovered, or redundancy through substitution (in the same way that lead pipes are no longer used for plumbing).

5.1.1 A case study—the lifetime of oil

To assess the true lifetimes of fuel reserves more accurately, we need to look at why and how R/P ratios change over time, and relate those changes to changes in both discovery and production. Figure 76 shows global oil reserves and oil production rates, crude oil spot price and R/P ratio for the period 1967–92. Since the Middle East is the world's main oil-producing region, some important events in Middle East history are also shown. Oil spot prices reflect world events closely.

This graph is quite difficult to interpret, so we will use another series of questions and answers to try and understand the data better. During our analysis, try to concentrate on the changes in slope of the curves in Figure 76 rather than on the values of the variables themselves.

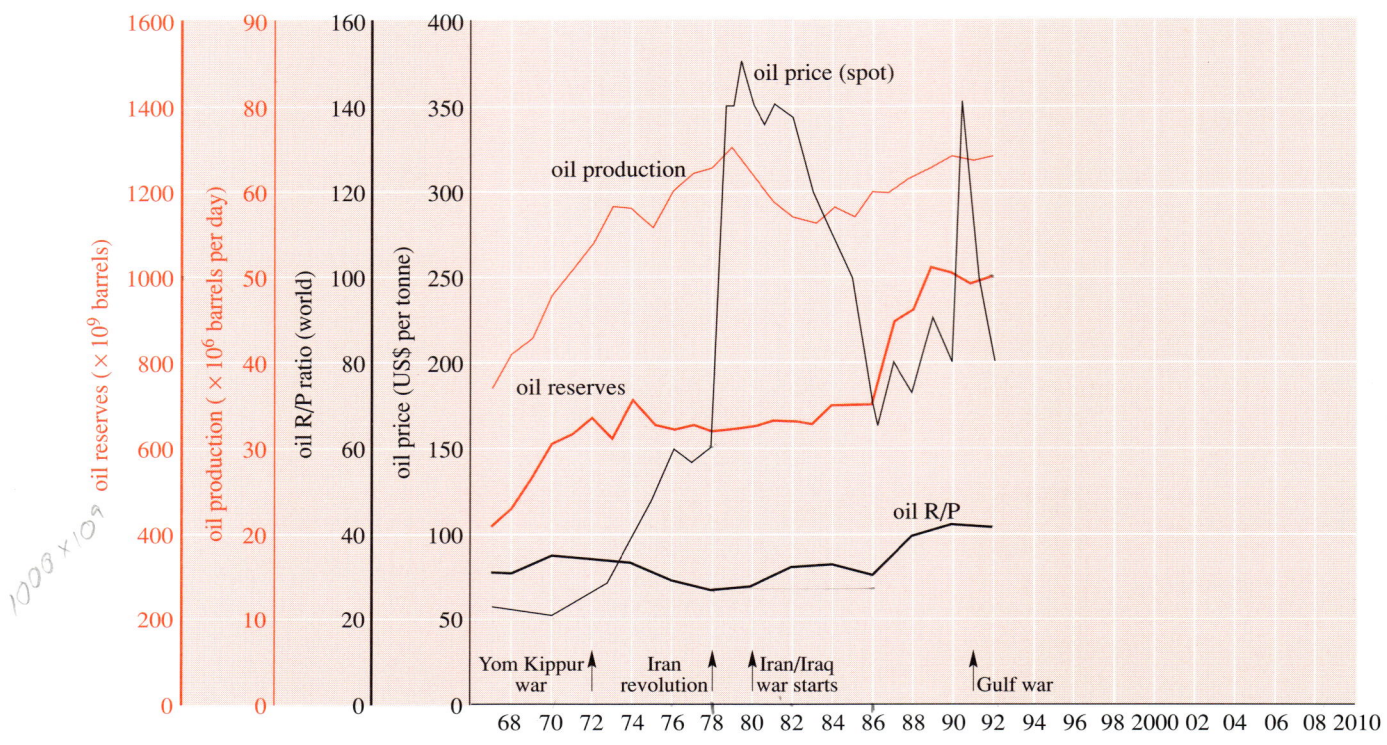

Figure 76 Variations in global oil discovery, production, price and R/P ratio over the period 1967–92. The four scales on the vertical axis of the graph relate to the four variables shown. Important political events in recent Middle East history are also shown.

- Describe briefly the trends in global oil *reserves* for 1967–92. How does the discovery pattern compare with that shown in Figure 65?

- Global oil reserves have generally increased since 1967, although for some brief periods (1972–73, 1974–75, 1989–91) oil reserves declined.

There were two periods of sustained increase in world oil reserves; 1968–72 and 1986–89. Changes in reserves do not seem to be closely related to major political events. The form of the curve, with sharp rises followed by periods of almost no increase in reserves, follows the theoretical discovery pattern shown in Figure 65.

- Describe briefly the trends in global oil *production* since 1967. What reasons could you suggest for that pattern?

○ Overall, global oil production has increased since 1967, but not in a regular way. There were significant falls in production in 1974, in 1979-83, in 1984 and in 1990. Each of these periods coincides with a period of political unrest in the Middle East.

From 1979 to 1983, reserves hardly increased, but over the same period production declined leading to an increase in R/P ratio — this is an example of a rise in R/P ratio driven by *falling production*. From 1986 to 1988, the rate of discovery of oil outstripped the rate of production — this is an example of a rise in R/P ratio driven by *rising reserves*. From this information we cannot tell why reserves have risen, whether it was because of new discoveries or because some former resources have for economic or political reasons become reclassified as reserves.

The price of a tonne of oil in US dollars in 1992 was some four times that in 1970, but allowing for inflation the real price of a tonne of oil is probably much the same as it was thirty years ago. However, there have been some spectacular fluctuations in oil price.

○ What factor appears to influence the price of oil most strongly?

○ Political disturbances in the Middle East.

A substantial rise in price followed the 1974 Yom Kippur war between Israel and its oil-producing neighbours. The two greatest leaps in oil price correlate almost exactly with significant changes in the relationship between one or more major oil producer and the industrialized West: the revolution in Iran in 1978 and the Gulf War in 1990. As the world adjusts to these political disturbances (usually by extra production elsewhere) the price of oil returns towards its 'base level'.

○ How much does oil price influence production?

○ Sharp rises in oil price soon lead to reduced production.

This is consistent with what we learned in Block 1, Section 2, where we saw that an increase in commodity price often produces a reduction in demand. High oil prices during 1979–82 were reflected in the decline of oil production over that period. Conversely, the fall in price during 1983–86 was accompanied by and followed by a steady increase in production. However, the industrialized world is so dependent on oil for energy and industry that substitution is not a practical short-term strategy, so price-driven drops in production are of limited duration.

○ How much does oil price influence reserves?

○ By and large, reserves are not directly affected by price.

This is rather surprising, since for other commodities a higher price has the effect of moving sub-economic resources into the category of reserves. The sharp rise in the price of oil in the mid-1970s stimulated exploration for oil in the North Sea and elsewhere, and may have been directly responsible for a significant increase in global oil reserves at that time. However, the rapid price rises in 1979 and 1990 did not lead to significant increases in reserves. The major increase in reserves in the late 1980s occurred at a time of relatively low prices, and followed a significant price fall.

We can make some general comments about the real lifetimes of oil, based on this case study. Over the latter half of the twentieth century oil production has been rising steadily, reflecting the developed world's need for oil to generate energy, provide transportable fuels, and be used as a **feedstock** for the petrochemical industry. Falls in production only occur when the oil price is high, through political unrest in the major oil-producing areas. The effects are generally short lived, but when they happen, they extend the lifetime of oil. Oil production will probably continue to rise into the future, driven not only by the continuing demand of the developed world but also by rising world population and increasing energy expectations and demands from the developing world.

Over the last few decades we have also been in a period of episodic sharp rises in oil reserves. We can expect that pattern to continue while there are new sedimentary basins to explore and new plays to be discovered. As long as new prospects exist, the oil R/P ratio can be expected to remain roughly constant. The limits of oil as a resource will not be appreciated until all the world's sedimentary basins have been fully explored.

The primary effect of the increased demand will be to drive the quest for discovery. With present demands and incentives, the world has a 'rolling cushion' of some thirty years of oil beyond the time when the limits of exploration are reached. If the high energy levels demanded by the developing world are met by increasing oil production, we can predict that the rolling cushion will drop below thirty years. If alternative energy sources and/or petrochemical feedstocks are found and used, the rolling cushion may well increase.

5.2 Video review: Energy at the Crossroads

Video Band 12: *Energy at the Crossroads* reviews some of the principal concepts covered in this Block, and links energy generation through fossil fuels with alternative ways of generating energy, which is the subject of Block 4 *Energy 2 – Nuclear and Alternatives*. Watch the video now; it will cement ideas and points from this part of the Block and focus your thinking on alternatives for fossil fuels.

Activity 6

In the latter half of Video Band 12, Norwegian politicians and company directors discuss their views on why the use of North Sea oil evolved the way it did, how best to develop the resource at present, and what the future holds for oil as an energy source.

Make two lists: one of the views (political and economic) of Rolf Magne Larsen and Arild Steine as representatives of a large oil company concerning the present state and future prospects for North Sea oil, and another of the corresponding views (political and economic) of Vidkun Hveding and Finn Lied as former Ministers in the Norwegian government.

Using the two lists, write brief synopses of the way North Sea oil might be developed in the future if (a) oil company directors or (b) politicians were to be in overall control.

Video Band 12 Energy at the Crossroads

Speakers

Hilary Irwin	Technical Advisor, Basin Modelling and Geochemistry, Statoil, Norway
Myles Bowen	Consultant and former Exploration Geologist, Shell UK
Colin Braithwaite	Head of Reservoir Engineering – Brent, Shell UK
Rt. Hon. Tony Benn MP	Secretary of State for Industry, 1974–75; Secretary of State for Energy, 1975–79
Vidkun Hveding	Minister of Petroleum and Energy, Norway, 1981–83
Arild Steine	Vice President – Public Affairs, Statoil, Norway
Rolf Magne Larsen	Executive Vice President – International Exploration and Production, Statoil, Norway
Finn Lied	Minister of Industry, Norway, 1971–72; Chairman, Statoil, 1974–83
Peter Sheldon	The Open University
Andrew Bell	The Open University

This video, which was made in 1993, has four main aims:

(a) It identifies some of the geological factors that contribute to oil and gas source rock maturation.

(b) It illustrates how oil exploration and production is planned and implemented.

(c) It contrasts electricity generation from fossil fuels with electricity generation from hydro power in Norway.

(d) It examines some of the scenarios for the transition from fossil fuels to renewable energy sources.

The following important points are made:

1. Hilary Irwin of Statoil, the Norwegian State Oil company, describes conditions during sedimentation of the Jurassic Kimmeridge shales. She shows how Cretaceous and Tertiary sedimentation in the Viking Graben in the North Sea has buried different parts of the Kimmeridge shales by different amounts, leading to both gas and oil generation.

2. Myles Bowen, the leader of the Shell/Esso exploration team that discovered the Brent oilfield in 1971, describes the reasons why Brent was a suitable target for oil exploration. Colin Braithwaite describes how oil and gas are produced from Brent and discusses the plans Shell have for changing from a waterflood production process to a reduced-pressure gas drive. He notes that this change should extend the life of the Brent field from 2004 to about 2015.

3. Tony Benn gives an account of the way in which negotiations were conducted between oil companies and the UK government during the 1970s.

4. Vidkun Hveding describes some of the value that self-sufficiency in hydroelectric power brings to Norway. The Norwegians have energy links with Denmark, enabling them to buy electricity from the Danes at night (preserving their own water supplies) and to sell electricity at peak times by increasing their own hydroelectric production.

5. Arild Steine and Rolf Magne Larsen discuss the future of oil reserves in the Norwegian sector of the North Sea from a company perspective.

6. Vidkun Hveding describes the problems of concentrating natural energy and notes that fossil fuels represent vast natural concentrations of early solar energy. In the long term, Hveding believes that nuclear power must be harnessed.

7. Finn Lied describes some of the alternatives to fossil fuels for energy generation and transport. Hydrogen-burning engines could provide clean, renewable power for transport. Lied believes that developed societies must use the 'window of opportunity' afforded by natural gas to clean up energy production from coal. Coal itself can then power electricity generation while solar power is developed.

5.3 Burning fossil fuels

The sort of analysis we made in Section 5.1 could equally well be made for other fossil fuels; the numbers would be different but the principles would be the same. However, it does make the assumption, also implicit in Video Band 12, that it is a good idea to continue to use fossil fuels the way we currently do. Is this assumption justified?

We may not get the opportunity to run out of fossil fuels. Instead, we may reach a limit where the undesirable effects on the natural system outweigh the benefits of generating energy from fossil fuels. What are these effects?

Fossil fuels are chemical compounds primarily made from carbon, hydrogen, nitrogen and sulphur, with traces of many other elements. What happens when we burn fuels, and what are the environmental consequences?

5.3.1 Burning pure hydrocarbons

Chemically, the simplest fossil fuel is the hydrocarbon methane, CH_4. Burning pure methane in oxygen (air) releases heat, which we then harness. Water and carbon dioxide are produced during the oxidation reaction. The basic chemical equation is straightforward:

$$CH_4 + 2O_2 = CO_2 + 2H_2O + \text{energy}$$

The amount of gases released by burning a given amount of methane can be calculated:

$$16 \text{ g of } CH_4 + 64 \text{ g of } O_2 = 44 \text{ g of } CO_2 + 36 \text{ g of } H_2O$$

So for every 1 kg of methane burned, 4 kg of oxygen is used, and 2.75 kg of carbon dioxide and 2.25 kg of water vapour are produced.

The reaction is similar if pure forms of more complex hydrocarbons are burned. For example, octane is C_8H_{18}. Pure octane burns according to the equation:

$$2C_8H_{18} + 25O_2 = 16CO_2 + 18H_2O + \text{energy}$$

For every 1 kg of octane burned, 3.52 kg of oxygen is used, and 3.10 kg of carbon dioxide and 1.42 kg of water vapour are produced. Burning the complex hydrocarbons that form petrol uses less oxygen and releases more carbon dioxide than burning the simple hydrocarbons found in natural gas.

Carbon dioxide and water vapour are gases that are emitted into the atmosphere when fuel is burned. On a cold day, the clouds that form when water vapour condenses into droplets can readily be seen coming from car exhausts. So burning hydrocarbons in air, whether it is in a gas-fired power station or in a car engine, automatically increases the amounts of water and carbon dioxide in the atmosphere, and changes the balance of the natural carbon cycle.

Where hydrocarbons are burned in a restricted oxygen supply, carbon monoxide (CO) can also be produced, and for this reason CO forms a significant part of car exhaust gases.

5.3.2 Burning the chemical impurities in fossil fuels

Theoretically, the only emissions from burning pure hydrocarbons should be carbon dioxide and water. But natural gas and crude oil are not pure hydrocarbons; they also contain sulphur, nitrogen and other impurities which also undergo chemical reactions when fossil fuels are burned.

Sulphur

Sulphur is a significant impurity in most fossil fuels, originating from the sulphur contained within the original decaying organisms. Virtually all the sulphur in fossil fuels oxidizes to sulphur dioxide (SO_2) during burning. The basic chemical reaction is:

$$S + O_2 = SO_2 + \text{energy}$$

Different forms of fossil fuel contain different amounts of sulphur. Natural gas, for example, contains virtually no sulphur; lignite has a sulphur content of around 1% by weight, compared with 2.5% for higher rank coal, 2.7% for orimulsion and 3.5% for fuel oil.

It is relatively difficult to remove sulphur from fuels before burning them. About half the sulphur in coal occurs as the iron-rich mineral pyrite (FeS), which could be separated from coal prior to burning by using a liquid of appropriate density to float off the light coal from the heavier pyrite. However, the size of the coal and the way which the pyrite is disseminated throughout the coal limit the effectiveness of this technique. Sulphur in coal that is not in the form of pyrite and the sulphur in petroleum are bound organically to the carbon molecules and cannot be removed in this way.

It is possible to remove sulphur dioxide from the waste gases after burning. SO_2 is removed from flue gases of power stations by reaction with limestone slurry in a *scrubbing unit*. The reaction that occurs (Block 2, Section 6.2) is:

$$2CaCO_3 + 2SO_2 + O_2 + 4H_2O = 2CaSO_4.2H_2O + 2CO_2$$
limestone $\qquad\qquad\qquad\qquad\quad$ gypsum

At the Drax power station in Yorkshire, which burns coal from the Selby coalfield, the limestone scrubbing unit removes 90% of the SO_2 from the flue gases. Gypsum useful to industry is produced by the process.

- ⚫ What are the implications for carbon dioxide release in using a limestone scrubbing unit to remove sulphur dioxide from power station flue gases?

- 🔴 Since CO_2 is also produced during the scrubbing reaction, more CO_2 is released into the atmosphere as a direct consequence of scrubbing than would be released if the gases were emitted direct.

On average, half the sulphur dioxide in the atmosphere comes from burning fossil fuels; the other half comes from natural causes, such as sea spray, rotting vegetation, plankton and volcanic activity. Region by region the situation is different; in the atmosphere over Europe about 85% of the SO_2 comes from burning fossil fuels.

Nitrogen

Nitrogen occurs in fossil fuels as part of the organic matrix and usually makes up about 2% of the fuel by weight. Nitrogen also makes up about 80% of the atmosphere, so it is always present when fossil fuels are burned in air. The two main oxides of nitrogen that form by burning fossil fuels in air are nitric oxide (NO) and nitrogen dioxide (NO_2), usually collectively known as NO_x.

NO_x emissions are the product of a complex series of chemical reactions which we need not detail, but high temperatures and high oxygen concentration favour NO_x production, whereas high carbon amounts and low oxygen concentrations favour the formation of nitrogen as a gas.

It is possible to reduce NO_x emissions either by burning under low oxygen concentrations, or by fitting catalytic reactors to reduce NO_x in flue gases to nitrogen. At present, the former treatment is more widespread than the latter.

Ash and particles

Crude oil and natural gas leave no ash when burned. Coal typically leaves at least 5–15% ash on burning. Orimulsion gives typically 0.3% ash residue. Ash from orimulsion combustion contains over 95% of metal compounds, so it has potential for metals recovery but equally carries the threat of metals pollution.

Ash particles and soot (carbon particles) can be emitted into the atmosphere through chimneys and exhaust pipes. Overall, less than 10% of the solid particles in the air come from burning fossil fuels. The vast bulk comes from volcanic eruptions, forest fires, erosion, the movement of dust by the wind and blowing salt spray.

5.4 Fossil fuels and atmospheric pollution

Water vapour, carbon dioxide, sulphur dioxide, nitrogen oxides, carbon monoxide, ash, soot and fuel particles are all released into the atmosphere when fossil fuels burn. The by-products of burning any given fossil fuel are similar whether the fuel is burned in power stations to generate electricity, burned to provide domestic heat or burned for transportation, but because the chemical and physical conditions are different for each of these applications, the by-products vary in amount from application to application. By-product emission varies from fuel to fuel, even when they are burned under similar conditions. To give one example, atmospheric smoke concentration in London has fallen tenfold since 1960 with the switch away from coal for home heating and electricity generation, but nowadays over 70% of the particulates in London air come from the exhausts of diesel-engined commercial vehicles.

Of the many by-products released into the atmosphere through burning fuels, water vapour is one of the most significant in terms of quantity, but since it is naturally part of the water cycle (as we saw in Block 3) there is as yet little concern over introducing more water vapour into the atmosphere. Three effects stemming from changing composition of atmospheric gases do currently cause concern; acid rain, ozone depletion and, most significantly, global warming through the greenhouse effect (which will be discussed in Section 5.4.3).

5.4.1 Sulphur dioxide and acid rain

The problems associated with SO_2 emissions are shown in Figure 77. They are serious and cross country boundaries. High chimneys provide ideal conditions for the oxidation of SO_2. Oxides and hydroxides of sulphur form in the atmosphere by the action of sunlight. These are turned by photochemical oxidization, helped by other gases such as ozone (O_3) and ammonia (NH_3) which act as catalysts, into first sulphurous acid (H_2SO_3) and then sulphuric acid (H_2SO_4). Depending on the amount of moisture in the air, up to 80% of emitted sulphur dioxide may be turned into these acids. Both

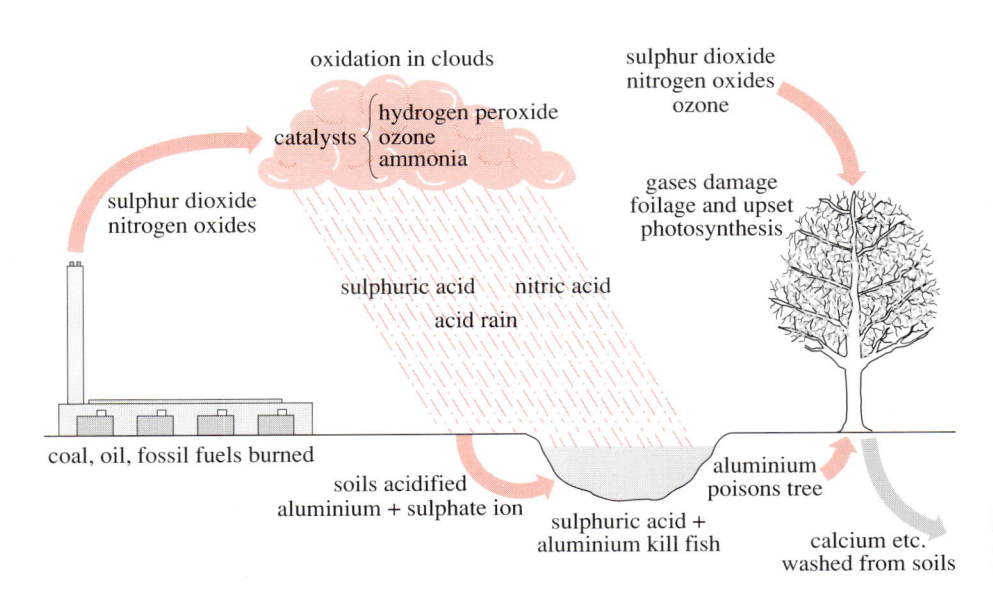

Figure 77 The chain of pollution caused by acid rain.

sulphurous and sulphuric acid dissolve in atmospheric water; the result is **acid rain**.

NO_x emissions also react with components of the atmosphere, contributing to acid rain by generating nitrous and nitric acid solutions. Acid rain increases the acidity of groundwater and can leach metallic ions from the soil. Some of these are poisonous to plants and animals (shown in Figure 77). Acid rain also attacks lime-rich soils and limestones exposed at the surface, releasing still more CO_2 into the atmosphere.

Acid rain is a product of burning fuels. Even if sulphur dioxide levels from power stations are strictly controlled, as they are in USA and several European countries, NO_x emissions from cars and power stations are capable of continuing the production of acid rain on about the same scale.

Current work suggests that even if SO_2 and NO_x pollution were halted today, the effects of acid rain outlined in Figure 77 would remain for decades. Acid and aluminium would continue to poison lakes and streams. One British investigation suggests that halving acid precipitation over the Scottish hills immediately would merely maintain the acidity of the local lochs at their present-day levels.

Smog

In December 1952 *smog*, a cold fog containing high levels of smoke and sulphur dioxide, hung over London for almost a week. It was one of the worst in a series of 'peasouper' fogs that descended on London at that time, and was directly responsible for about 4000 deaths through bronchial infections and heart attacks.

At the time, doctors blamed high smoke levels in the fog, but more recent research suggests that the formation of highly acid particles may have been important. The 1952 London smog had a pH of 1.6, more acid than lemon juice. A public outcry led to smoke controls and a wider use of smokeless solid fuel.

The advent of North Sea gas and the siting of power stations outside the centres of population mean that smoke-laden fogs are rare nowadays. However, they have been replaced by a different form of air pollution; NO_x and smoke particles from vehicle emissions. A smog that built up from London traffic fumes during four windless days in December 1991 was the worst in Britain in recent years, and directly responsible for the deaths of 160 people. Two pollutants were exceptionally concentrated in the London air: NO_2 levels reached 423 parts per billion, the highest level ever recorded in Britain, and black smoke particles reached 228 $\mu g\,m^{-3}$.

Block 4

Activity 7

Figure 78 shows the pattern of atmospheric fallout of sulphur (mainly as sulphates) over Europe that was typical in the late 1980s. Use information in this Block, from the previous parts of this Course, and from a European atlas, to analyse Figure 78. Try to answer as many of these questions as possible:

(a) List the countries which have the highest fallout of sulphur (over 6 g m^{-2}), and the countries which have the lowest fallout (under 0.8 g m^{-2}).

(b) Giving some thought to the geology of Europe, what evidence is there in Figures 39, 41 and 79 to show that sulphur precipitation is largely caused by solid fuel power stations?

(c) Why should Ireland have such a low fallout, despite generating much of its electricity from solid fuel, and Denmark and southern Sweden have a high fallout despite having no coal?

(d) Suggest a strategy for reducing sulphur precipitation in Europe, stating which countries need to take action and what that action might be.

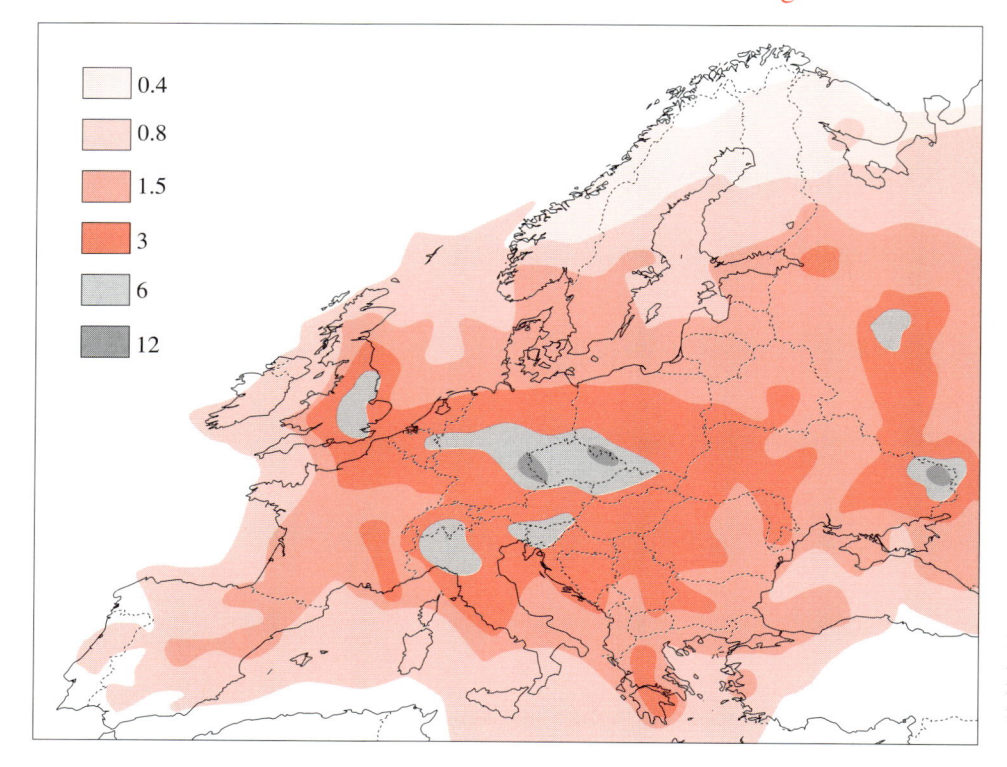

Figure 78 The typical fallout of sulphur, in grams per square metre, over Europe in one year in the late 1980s.

5.4.2 Ozone

Ozone (O_3) is a form of oxygen which has three oxygen atoms in each molecule rather than the more familiar two (O_2) which make up the kind of oxygen we breathe. Ordinary oxygen molecules are broken apart by certain wavelengths of ultraviolet radiation into highly reactive O atoms. Some of these will react with other O_2 molecules to form ozone. This reaction takes place in the stratosphere (the upper part of the atmosphere), and has the effect of absorbing ultraviolet light from the Sun. The stratosphere is heated as a consequence.

Ozone itself reacts under the influence of longer wavelengths of ultraviolet light, breaking down into O atoms and O_2 molecules. Most of the single oxygen atoms produced combine with other oxygen molecules to re-form ozone. The composition of gases in the stratosphere is unchanged, but more ultraviolet light has been absorbed and the stratosphere has become

Car exhausts in Mexico City

Mexico City has the worst car exhaust problem in the world. Cars there are typically Volkswagen beetles, several years old. They run on unleaded petrol and cannot therefore use catalytic converters. The city lies at an altitude of 2400 m where the air is thin, so the carburettors of many of the cars have been adjusted to enrich the fuel–air mixture. The engines of these cars run on an air–fuel ratio of 8 : 1. From an emissions point of view, this compares unfavourably with both the 12 : 1 ratio which enables engines develop peak power at sea level, and the 19 : 1 ratio which is the best overall for cutting exhaust emissions.

Because the fuel mixture is so rich, much of the petroleum passes through the engine unburned, and the air in Mexico City is virtually saturated with hydrocarbons from buses, cars and trucks. To counter this, some cars are now being fitted with a computerized sensor which monitors the rotation of the engine's flywheel and reduces the richness of the fuel–air mixture to as low a level as possible. In effect, the microchip keeps the engine on as lean a diet as possible without compromising power.

This microchip can reduce carbon monoxide in the exhaust by 60–70%, total hydrocarbons by 20–40% and nitrogen oxides by some 20%. Fuel efficiency is improved by 10–15%. The technology can work on modern fuel-injection engines as well as older engines with carburettors. Why haven't we all got one?

---

warmer. Ozone is therefore being constantly produced and destroyed and the natural system is in dynamic balance.

Other atoms and molecules, notably NO_x and chlorine (Cl), destroy ozone more permanently:

$NO + O_3 = NO_2 + O_2$ $\qquad$ $Cl + O_3 = ClO + O_2$

$NO_2 + O = NO + O_2$ $\qquad$ $ClO + O = Cl + O_2$

Nitric oxide formed at the end of one cycle can combine with more ozone before eventually being converted through NO_2 into nitric acid. Similarly, chlorine can be recycled many times. Both reactions have the effect of reducing ozone molecules to oxygen molecules. Much of the atmospheric NO_x comes from burning fossil fuels, and the atmospheric chlorine from breakdown of chlorofluorocarbons (CFCs) under ultraviolet light.

Stratospheric ozone absorbs ultraviolet radiation and prevents it from reaching the ground, where it could cause sunburn, some forms of skin cancer and eye problems such as cataracts. However, ozone is a hazard close to the ground. It can damage plants and materials from rubber to textiles; it can trigger asthma and bronchitis and it enhances the formation of acid rain. Low-level ozone is formed from photochemical reactions involving both NO_x and traces of hydrocarbons in the air. Much of both comes from car exhausts and power station flues, but methane is often introduced through natural causes such as the flatulence of cows. Low-level ozone might best be controlled, along with acid rain, by tackling nitrogen oxide pollution.

5.4.3 Carbon dioxide, methane and global warming

We saw in Section 5.3.1 that carbon dioxide is the most abundant waste product of burning fuels, but the amount of carbon dioxide produced varies from fuel to fuel. Natural gas produces the least carbon dioxide. Burning a tonne of natural gas in a power station releases about 2.75 tonnes of CO_2; burning a tonne of petrol in a car releases over 3 tonnes of CO_2. Orimulsion and fuel oil produce about the same amount of carbon dioxide as each other when burned. Coal produces about 20% more, since coal is composed of more complex carbon molecules which yield more carbon dioxide per tonne than simple hydrocarbons like methane.

In the UK in 1990, burning fuels generated 160 million tonnes of CO_2, just under 3 tonnes per person per year. Of this, 42% came from burning coal, 35% in total from burning liquid petroleum and 19% from burning natural gas (Figure 79). Worldwide, $5–6 \times 10^9$ tonnes of carbon are released into the atmosphere every year through burning fossil fuels.

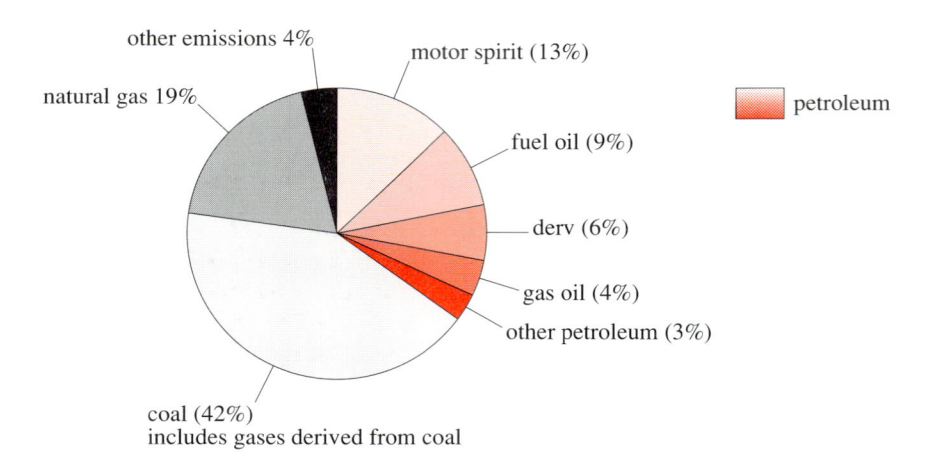

Figure 79 Percentage carbon dioxide emissions in UK in 1990, by type of fuel.

- How does the yearly amount of carbon released through burning fossil fuels compare, as a percentage, with that fixed by, say, land plants?

- Figure 6 showed that 110×10^9 tonnes of carbon is fixed by plants each year. The carbon flux from burning fossil fuels is approximately 5% of this.

- How will this carbon flux affect the natural carbon cycle?

- We saw in Section 2 that there was roughly a net balance of natural carbon fluxes between the atmosphere and the terrestrial and marine stores. Burning fossil fuels adds roughly a 5% flux of carbon into the atmosphere.

Also, between 4×10^7 and 10×10^7 tonnes of methane is released into the atmosphere each year as a result of winning fossil fuels, through gas exploration and transmission and during coal mining.

Until recently, the atmosphere was considered to be large enough to absorb large quantities of carbon-based gases and dilute them to harmless levels. We now believe that this is not the case. Rapid increase in carbon-based gases is interfering with the energy budget of the Earth itself by enhancing the **greenhouse effect** (see Box).

Some greenhouse gases, such as methane and CFCs (chlorofluorocarbons), are far more potent absorbers of longer wavelength radiation, molecule for molecule, than carbon dioxide and water. Some gases are more unstable in the atmosphere than others; methane, for example, has only a relatively short lifetime (around ten years) before it breaks down into carbon dioxide and water. Each gas therefore has its own warming potential, based on its potency and lifetime. One kilogram of methane released into the atmosphere now can potentially cause eleven times more warming over the next hundred years than one kilogram of CO_2. The most threatening additions to our atmosphere because of their potency are methane, NO_x and the CFC-related gases, and concern is being expressed about the large volumes of atmospheric CO_2 added through human activities.

The greenhouse effect

All hot objects emit electromagnetic radiation. The wavelength of radiation that they emit depends on their temperature — the hotter any object is, the shorter the wavelength of emitted radiation. The temperature of the Sun is over 5500 °C, so it emits short-wavelength radiation (Figure 80) in the ultraviolet, visible and near infrared parts of the spectrum. Much of the Sun's radiation is visible light.

About 30% of the incoming, short-wavelength radiation from the Sun is reflected back out into space from clouds, from the outer layers of the atmosphere and from the Earth's surface. The remaining two-thirds is absorbed either by the ozone-rich upper atmosphere or by materials at the Earth's surface. Absorbed radiation would tend to heat up the planet, but a dynamic energy balance is maintained because the Earth is itself hot, and re-radiates energy back towards space. But the Earth is much cooler than the Sun, so terrestrial radiation is emitted at longer wavelengths than solar radiation.

Most atmospheric gases absorb radiation. The specific wavelengths absorbed by each component gas depends on its composition and on its concentration. For example, we saw in Section 5.4.2 that oxygen and ozone effectively absorb ultraviolet solar radiation. The greater the concentration of ozone in the upper atmosphere, the more ultraviolet radiation it absorbs. The parts of the spectrum absorbed by six of the more significant atmospheric gases is shown in Figure 80.

The atmosphere is more or less transparent to many wavelengths of incoming solar radiation, particularly in the visible part of the spectrum. However water, carbon dioxide, methane, nitrous oxide (N_2O) and ozone each absorbs wavelengths throughout some part of the infrared range emitted by the Earth. Taken together, atmospheric gases absorb most wavelengths of terrestrial radiation, but water and carbon dioxide make the most significant contribution. Nitrogen, which makes up 80% of the atmosphere, is an exception; it does not absorb any infrared radiation.

The energy acquired by molecules of atmospheric gases when they absorb radiation gives rise to an increase in their temperature. Since almost all the mass of the atmosphere lies within 30 km of the Earth's surface, and half within the lowermost 6 km, the heating effects are felt most closely at the lowest atmospheric levels. In all, the Earth's surface is some 33 °C warmer than it would be if it had no atmosphere. This phenomenon is the *greenhouse effect*, and without it the Earth would be in a permanently ice-bound state.

There is a natural moderating effect to this 'greenhouse' heating. A warmer atmosphere is less able to absorb longer wavelength radiation, so as the atmosphere warms it becomes less effective at insulating and tends to cool. The reverse is also true; a cooler atmosphere absorbs more energy and heats up. The Earth's heat gains and losses are best balanced if the temperature of the atmosphere remains at a high mean value of 15 °C.

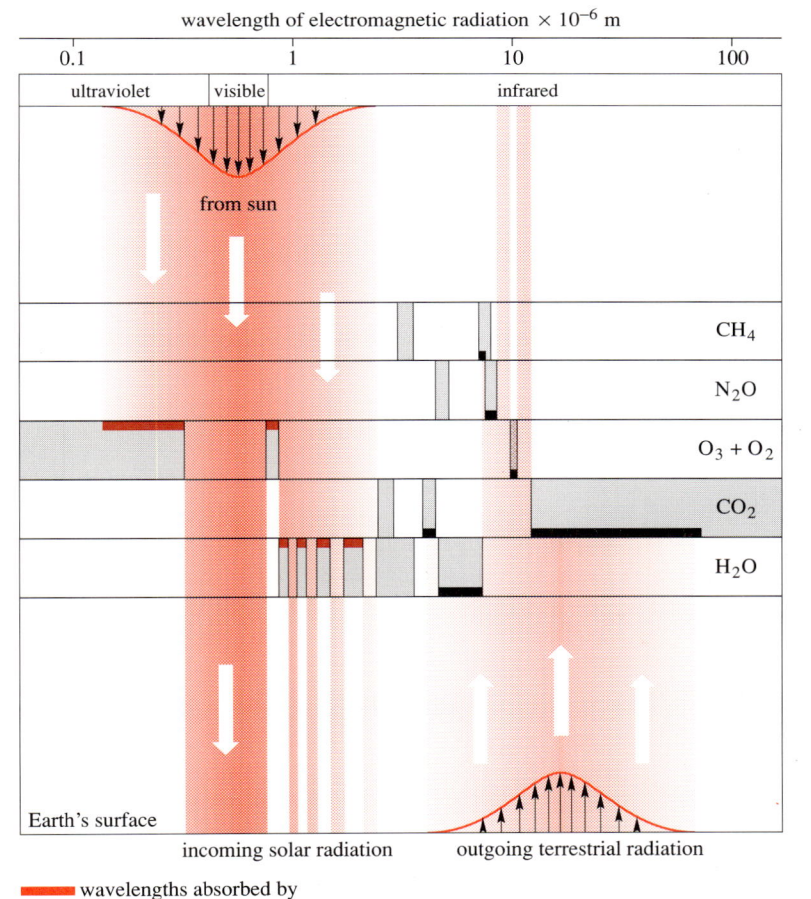

Figure 80 The origin of the natural greenhouse effect. Short-wavelength energy from the Sun is absorbed by oxygen, ozone and water vapour but most wavelengths reach the Earth's surface. Much longer wavelength radiated from the Earth is largely absorbed by atmospheric gases, causing a heating effect.

Global warming

Over the last two hundred years, human industrial and agricultural activity has caused a change in the concentrations of atmospheric gases. In particular, there have been dramatic increases in the production of carbon dioxide and methane by industry, energy production and modern transport. CFCs have appeared in the atmosphere as a result of modern industrial processes. Over this time the world has been getting warmer; there has been an increase in the *global mean surface temperature* of more than 0.5 °C.

Global warming seems to be having a more significant effect on night-time rather than daytime temperatures. Since 1950 the minimum daily temperature over most of the landmass of the northern hemisphere has risen three times as fast as the maximum temperature. In the UK, nights are on average 0.84 °C warmer than they were in 1950, whereas days are only 0.28 °C warmer; the trend applies to all northern continents and all seasons.

In North America and Europe, cloud cover has increased along with the warmer nights. Low-altitude clouds limit the loss of heat from radiation at night and shade the ground from sunlight, limiting the warming effect by day. Two causes have been postulated for an increase in cloud cover. One is pollution, since polluted air tends to encourage the formation of denser and more numerous clouds. The other is ocean evaporation; an enhanced greenhouse effect may be increasing evaporation from the oceans, leading to increased cloud cover over land.

The effects of changing atmospheric CO_2 concentrations are clearly preserved in polar ice bubbles. During polar precipitation, air is trapped by falling snow and preserved as bubbles in the uppermost layers of the icecap. Where snowfall has accumulated over time, layers many thousands of years old lie buried under successively younger layers. Compacted snow turns to ice, and preserves within it bubbles that record the chemistry of the atmosphere as it was when the snow fell. These bubbles can be analysed to yield amounts of atmospheric gases, including CO_2. The temperature and age at which the bubble formed can also be determined from isotope studies.

Aerosol cooling

There is growing evidence that global warming has been put into reverse above some populous countries of the northern hemisphere. The effect is caused by increases in levels of sulphate and dust particles (*aerosols*) from power stations in Europe, dust storms in the Sahara, farmers burning tropical rainforests and Far Eastern iron and steel manufacturers.

Aerosols cool the atmosphere in two ways. They reflect and scatter sunlight, slightly reducing the amount that reaches the Earth. Sulphate aerosols also make good nuclei for condensation of water vapour, so they encourage the formation of clouds. Large numbers of aerosols produce many small water droplets, which then form whiter, more reflecting clouds. Clouds shade the ground during hot summer days, but at night and in winter they warm the surface layers of the atmosphere by absorbing long-wavelength radiation emitted from the Earth's surface.

In the USA, average daytime cloud cover has risen from just less than 50% in 1900–40 to above 58% since 1960, and whereas the average daytime temperature of much of the world's land mass has risen since 1950, the areas with high SO_2 emissions have cooled. Maximum daytime temperatures between June and November, averaged over the land areas of the northern hemisphere for the past 40 years, have fallen by 0.4 °C.

Cloud cooling is concentrated near the sources of air pollution in middle and high latitudes of the northern hemisphere, increasing the temperature difference between these regions and the tropics. This is the opposite to predictions made by some 'greenhouse' models, but fits well with temperature changes that have actually been observed over past decades.

Whereas the effect of greenhouse gases is global, the effect of aerosol cooling is local. Some scientists believe that aerosol cooling can completely mask greenhouse warming over the eastern USA and central Europe. The greatest cooling effect at present is over Eastern Europe which, at 4.2 W m^{-2}, is twice the current greenhouse warming effect there.

Figure 81 plots CO_2 concentration against the change in atmospheric temperature, compiled from information contained in an ice core drilled at Vostock, Antarctica. For over 160 000 years (the age of the lowest part of the drill cores) temperature changes have in general matched CO_2 concentration. For example, temperature peaks at about 130 000 years and 10 000 years ago are accompanied by increased CO_2 concentrations, whereas temperature lows at 65 000 years and 20 000 years are accompanied by decreased CO_2 concentrations.

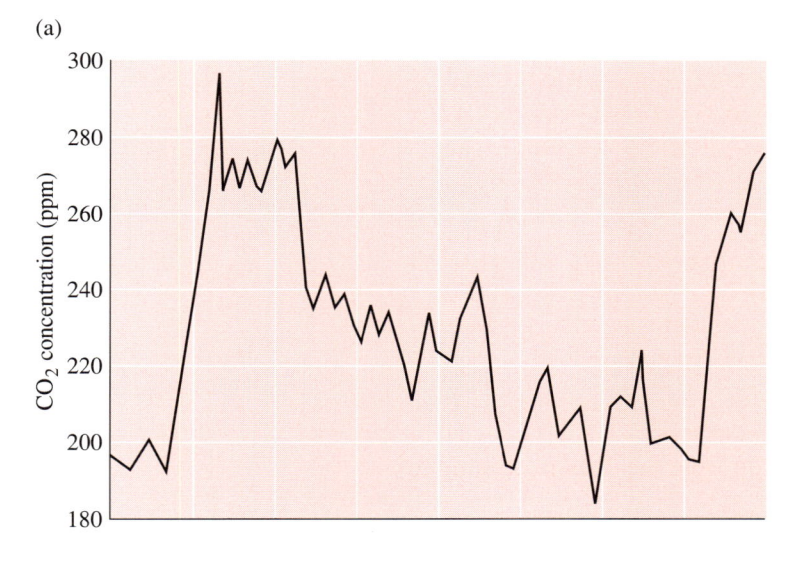

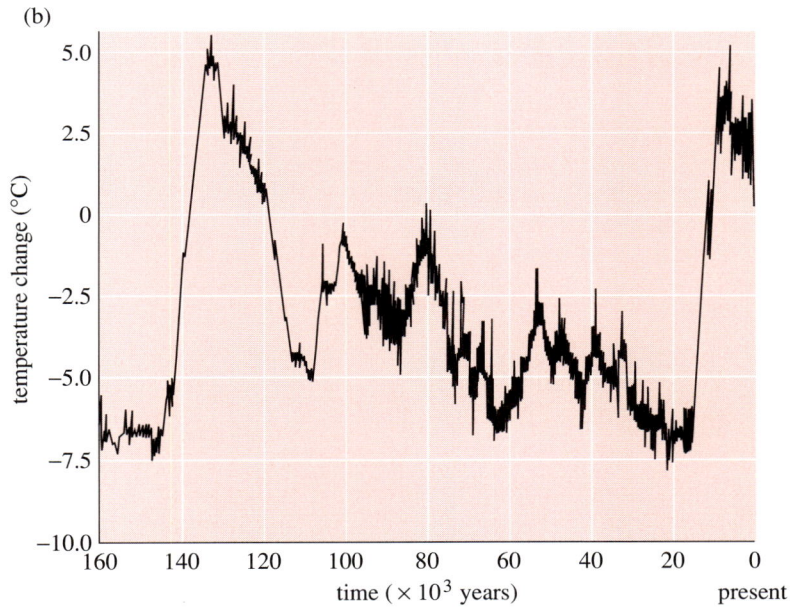

Figure 81 (a) The record of CO_2 concentration in parts per million and (b) temperature change revealed by cores taken from the Antarctic icecap.

- Does Figure 81 prove that increases in atmospheric CO_2 leads to overall temperature rises?

- No, we cannot tell from this information whether high CO_2 levels lead to warmer temperatures or warmer temperatures cause an increase in CO_2 concentration.

Figure 81 provides good evidence that global temperatures and CO_2 concentrations are linked. As yet, it remains unproven that recent increases in atmospheric CO_2 and other greenhouse gases are the direct cause of global warming. However, human activities have added sharply to the current amount of CO_2, such that global concentrations reached 355 p.p.m. in 1992 and are expected to reach 600 p.p.m. in the next 40 years. These values are twice the maximum recorded in the Antarctic core. The implication is that sharp increases in CO_2 concentration in the next few decades could lead to significant temperature rises.

Much of the fossil fuel resource that took many millions of years to accumulate has been burned in the 200 years since the Industrial Revolution. This sudden return of carbon dioxide and water into the atmosphere gives rise to much environmental concern through enhanced global warming. A 'business-as-usual' scenario (no change in approach from what we are currently doing) of industrial greenhouse gas emissions would, by 2028, result in an equivalent doubling of the pre-industrial carbon dioxide level. The mean global surface air temperature would increase at about 0.3 °C per decade; faster than any rise seen over the last 10 000 years. By 2030, temperatures would be 0.7–2.0 °C higher than present; by the end of the twenty-first century the rise could be 3 °C.

These increased temperatures are likely to be accompanied by a rise in global sea level, because of thermal expansion of the oceans and melting of the polar ice-caps. The 'business-as-usual' scenario suggests a sea-level rise of 20 cm by 2030 and 65 cm by the end of the twenty-first century.

To stabilize CO_2 at present-day levels would require an immediate reduction of emissions from human activities of over 60%. The message is clear; to reduce global warming we must control the production and emission of greenhouse gases, particularly CO_2. Most of all this means finding alternative sources of energy for transport and for electricity generation.

5.5 What about the alternatives?

We have devoted the whole of this block so far to fossil fuels because they provide the overwhelming part of humanity's enormous energy needs. You have seen that reserves of these fuels are sufficient to last for several decades (perhaps a couple of centuries in the case of coal). However, as we showed in this last Section, it is widely held that extraction and use of fossil fuels is polluting the environment, in particular through our addiction to petroleum-driven motor vehicles and aircraft.

At the 1992 Earth Summit at Rio de Janeiro, it was agreed that the major industrialized nations would stabilize their CO_2 emissions from fossil fuels at *1990 levels* by the year 2000. This decision was taken primarily to relieve possible global warming caused by any greenhouse effect, but such a policy would of course also stabilize emissions of other major contributors to atmospheric pollution and acid rain, such as oxides of sulphur and nitrogen.

Our own energy consumption doubled between 1960 and 1990. Will it go on increasing at that rate? If so, we cannot possibly hope even to reduce our emissions of CO_2, let alone stabilize them at 1990 levels, unless we use our fossil fuel resources much more efficiently and sparingly.

Perhaps the best policy would be to turn to some resource other than fossil fuels to allow us to maintain present levels of energy consumption yet reduce emissions. What options are available to us? Can we obtain energy, particularly electricity, in an effective way from resources other than fossil fuels? Clearly, any alternatives cannot have the same benefits to us as fossil fuels, otherwise we would not have become so addicted to them. Some energy analysts such as Finn Lied (Video Band 12: *Energy at the Crossroads*) urge that we are already late in making the change. But do we really need to change now, bearing in mind that economic reserves of oil and gas will last for several more decades even at present rates of extraction?

Anyway, what do we want all that energy for? An alternative view is that we should reduce our use of fossil fuels by reducing our energy consumption rapidly and significantly. That means we would have to adopt alternative ways of doing things: greater energy conservation at home and at work, less travel overall, better use of mass transport, different life styles and expectations, and so on.

We said at the beginning of this Block that by the year 2100 the world could be using twice as much energy as today. This is a mind-boggling statistic, made all the more daunting because the estimate has been made assuming that the average energy use per capita would be only the same or even less than it is today. The sharp increase is simply because the world's population is by then expected to be at least double what it is now. Even if we make major savings through efficiency, we will still use a prodigious amount of energy. Where can it all come from?

Questions such as these are addressed in the second part of Block 4 *Energy 2 – Nuclear and Alternatives*, in which we consider the nuclear and alternative options for energy that are presently available to us, and examine the extent to which they could play a more important role in future energy provision than they do at present.

5.6 Summary of Section 5

1. The R/P ratio does not vary simply year on year. To assess lifetimes of fuel reserves more accurately, we need to look at changes in R/P ratios over time, and relate those changes to the factors that moderate discovery and production.
2. As long as new prospects exist, petroleum R/P ratios are expected to remain roughly constant. The limits of petroleum resources will not be reached until all the world's sedimentary basins have been fully explored. With present demands and incentives, the world has a 'rolling cushion' of some thirty years of oil beyond the time-limits of exploration.
3. Burning pure hydrocarbons in air releases heat, water and carbon dioxide. The more complex hydrocarbons found in petrol produce more CO_2 and less water than the simple ones found in natural gas.
4. Impurities such as sulphur, nitrogen and ash occur in varying amounts in fossil fuels. Burning the fuel also liberates sulphur dioxide and nitrogen oxides.
5. Three effects of burning fossil fuels currently cause much environmental concern: acid rain, ozone depletion and global warming through enhancing the greenhouse effect.
6. By the action of sunlight and catalysts, SO_2 is turned into first sulphurous acid (HSO_3) and then sulphuric acid (H_2SO_4), which dissolve in atmospheric water to produce acid rain. NO_x emissions also react with components of the atmosphere, and form acid rain by generating nitrous and nitric acid solutions. Acid rain damages plants and animals, and poisons water.
7. Ozone is naturally produced by the action of sunlight on oxygen in the upper atmosphere. Ozone formation and destruction are parts of a naturally balanced cycle. Stratospheric ozone absorbs ultraviolet radiation and prevents it from reaching the ground, where it could cause sunburn, skin cancer and eye problems. NO_x from burning fossil fuels and chlorine can destroy ozone permanently. Ozone close to the ground forms from the action of light on car exhausts, and can damage plants, trigger asthma and bronchitis, and enhance the formation of acid rain.

8   There is a natural greenhouse effect that already keeps the Earth's surface 33 °C warmer than it might otherwise be. Comparatively high atmospheric concentrations of waste gases, notably CO_2, from burning fossil fuels and other human activity are enhancing the greenhouse effect and appear to be causing global warming. An immediate reduction of CO_2 emissions from human activities of over 60% is required if we are to stabilize atmospheric CO_2 at present-day levels.

Questions 33 to 36

Each of the Questions below consists of two statements (a) and (b), linked by the word *because*. Study each pair of statements carefully and, for each pair, decide which of A–E in the Key below is most appropriate:

Key

A   Both statements (a) and (b) are incorrect.
B   Statement (a) is correct but statement (b) is incorrect.
C   Statement (b) is incorrect but statement (a) is correct.
D   Both statements are correct, but statement (b) *does not* provide a valid explanation for statement (a).
E   Both statements are correct, and statement (b) *does* provide a valid explanation for statement (a).

Question 33

(a) The current R/P ratio of any fossil fuel is not a good indicator of the lifetime of that fuel

because

(b) the lifetime of any fossil fuel is heavily dependent on future discovery and production rates which are not reflected in the current R/P ratio.

Question 34

(a) Global oil reserves increased between 1986 and 1989 from 7×10^{11} barrels to over 10×10^{11} barrels

because

(b) prior to 1986 there had been a twelve-year period of relatively minor changes in total global oil reserves.

Question 35

(a) Acid rain will necessarily increase as the energy requirements of the developing world get larger

because

(b) acid rain is caused by the photochemical oxidation of sulphur dioxide and nitrogen oxides, which themselves largely come from burning fossil fuels.

Question 36

(a) Global warming is definitely caused by burning fossil fuels

because

(b) the heat that escapes from power station cooling towers and car exhausts is causing a greenhouse effect which is warming the atmosphere.

OBJECTIVES

Now that you have completed this Block, you should be able to do the following.

1. Explain in your own words, and use correctly, the terms in bold introduced in this Block and listed in the *Glossary*.
2. Understand the difference between energy and power, and state roughly how much energy different societies typically demand each year.
3. State the relative contributions of different natural energy sources to the Earth's energy budget, and discuss some of the problems associated with concentrating, storing and transporting energy.
4. Distinguish between renewable and non-renewable energy resources.
5. Outline why some rocks are sources of fossil fuels whereas others are not, recognize the geological factors that are likely to generate carbon-rich rocks and predict which carbon-rich rocks would be likely to source oil and/or gas.
6. List the types of geophysical data that can be obtained by drilling boreholes, and show how these can be used to confirm the suitability of the rock sequence as a host for coal.
7. Identify the main geological factors that limit production of underground and opencast coal, and recognize the ways that these factors are assessed and overcome.
8. Identify the main coal-bearing strata in UK, locate the areas of substantial existing UK coal reserves, and locate the areas of substantial existing coal reserves worldwide.
9. Classify geological structures that can trap oil or gas, and give examples of the important types of oil and gas source rocks, reservoirs and traps.
10. Explain why seismic surveys are the most appropriate remote sensing technique for locating oil- or gas-rich horizons in the subsurface, and outline briefly the techniques used to produce oil and gas from onshore and offshore wells.
11. Understand the environmental effects of transporting and storing fossil fuels.
12. Identify the main UK oil and gas plays, locate the areas of substantial existing UK oil and gas reserves, and identify some of the main plays in the rest of the world.
13. List and describe the unconventional sources of petroleum.
14. Estimate the lifetime of known fossil fuel reserves at given rates of discovery and use, and explain where and why significant reserves of conventional fossil fuels may be found in the future.
15. Describe the chemical products of burning fossil fuels, and recognize the effects of increased amounts of these products on the natural environment.

ANSWERS TO QUESTIONS

Question 1

(a) It's almost impossible that any of the objects around you required no energy at all in their formation. All manufactured items, the pages of this Block, the walls of the room etc., require energy for their production.

(b) You may be looking at some things that required no *human-made* energy in their formation, such as flowers and plants, a piece of fresh fruit, the family pet, or your rocks and minerals collection.

Question 2

The two 2 kW electric fires on for 12 hours use

$(2 \times 2000 \times 60 \times 60 \times 12)$ J = 172.8 MJ of energy

The 650 W microwave used for 1.5 hrs needs

$(650 \times 60 \times 60 \times 1.5)$ J = 3.5 MJ of energy

The 500 W fridge, running all day, needs

$(500 \times 60 \times 60 \times 24)$ J = 43.2 MJ of energy

The shower needs

$(5000 \times 60 \times 60)$ J = 18 MJ of energy

The washing machine needs half the energy of the shower

9 MJ each day

Lights use

$(2 \times 150 \times 60 \times 60 \times 24)$ J = 25.9 MJ each day

(a) The daily energy requirement is therefore

$(172.8 + 3.5 + 43.2 + 18 + 9 + 25.9)$ MJ = 272.4 MJ every day

(b) The average power requirement is

$272.4/(60 \times 60 \times 24)$ MJ per second, or 3.2 kW

Question 3

Your additions to Figure 2 should look like Figure 82 below.

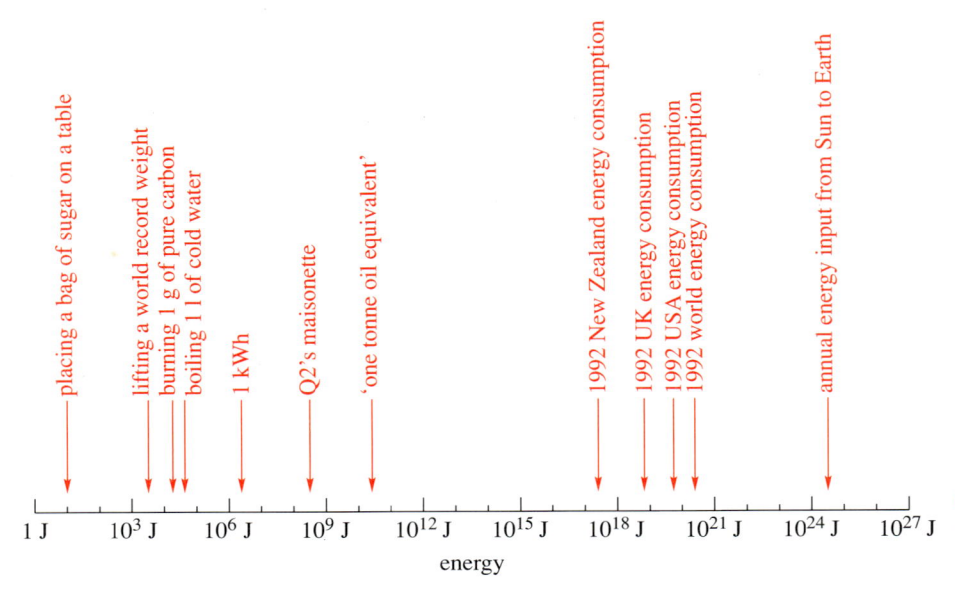

Figure 82 A fully completed version of Figure 2 for use with Questions 3, 4 and 5.

Answers to Questions

Question 4

We know that the Sun supplies 17×10^{16} W to the Earth, of which 10×10^{16} W enters the Earth's system (atmospheric heating, 2×10^{16}, + absorbed at surface, 8×10^{16}). 10×10^{16} W is equivalent to 10^{17} joules per second, and since there are 3.15×10^7 seconds in a year, the Sun's annual energy supply is

$$3.15 \times 10^7 \times 10^{17} \, \text{J} = 3.15 \times 10^{24} \, \text{J}$$

Question 5

Statements (a)–(e), and (h) are correct. Statements (f) and (g) are incorrect. In (f), the average global power demand in the 1990s is expected to be around 10 TW, but this figure is expected to rise to 20–50 TW in the following century as world population rises and currently underdeveloped countries require more energy for industrialization. In (g), the heat flow up through the surface from the Earth's interior is only roughly one-five-thousandth of the solar energy that reaches the Earth's surface.

Question 6

A (e); B (d); C (b); D (c); E (a).

Question 7

(a) Coal is made from (i) bright woody material (called vitrinite on the video), formed from plant or bark material, (ii) films of vitrinite in a dark matrix, formed from strips of woody material in a matrix of spores; (iii) dull hard material (called 'inertinite' on the video), formed from resin; and (iv) dirty friable material, formed from crushed woody material.

(b) In the video, where the seams contain no carbonates, the two most important impurities are identified as 'dirt': shale or mud deposited within the seam, and the mineral pyrite. The mudrock does not burn but scorches to becomes ash, whereas on heating the pyrite oxidizes to release the gas sulphur dioxide.

(c) In a 'typical' coal cycle, the sequence is coal — marine band — muddy sediment (shale) — sandstone soil (or seatearth) — next coal.

(d) A washout is a lobe of sand which was deposited into a channel and which eroded away the seam below. Washouts often form as breaches in the banks of a river during flooding. Washouts are simply avoided when opencast mining, but cause roof problems in deep mines.

(e) Coal swamps are typically extensive, flat-lying plains at or near sea level. They are necessarily densely vegetated. Swamps like this existed in late Carboniferous times, about 300 Ma ago, in Britain.

(f) Peat is turned into coal by heat and pressure, due to compaction and burial by overlying sediments.

Question 8

(a) None of them did, since they are all missing from Figure 15. Iceland formed long after the North Atlantic Ocean opened, itself some 70 Ma after the Upper Jurassic. The Bay of Biscay also formed somewhat later.

(b) Northern Denmark and England; the rest were already land areas in Kimmeridge Formation times, so no marine sediment could have been deposited there.

Question 9

(a) Figure 17 shows that these rocks have not been heated sufficiently for oil to form, so only biogenic methane will be present (Figure 16). Note that a slightly higher temperature (at least 65 °C) would have produced some oil.

(b) The high temperature means that natural gas has been generated.

(c) This rock has been subjected to such high temperature that the oil has been destroyed and the gas has been lost.

Question 10

(a) The Green River oil shales are thinly laminated fine-grained rocks some 80 m thick. They are composed of thin layers (laminae) of alternating brown muddy bands rich in organic matter and paler silty quartz bands. In all, they contain about 15% carbon, mostly long-chain hydrocarbons.

They were thought to originate in an inland lake. The layers represent seasonal variations during the year; the brown organic layers form from plant life which flourished around the lake in hot seasons, whereas the paler layers represent sediment washed into the lake during rainy seasons.

(b) Geokinetics are exploiting a flat-lying bed of oil shales lying close to the land surface. First, holes are drilled into the oil shale; deeper at one end of the area than the other. Then explosives are detonated at the bottom of each hole simultaneously to open fractures and make the oil shale more permeable. Air is pumped into the shallower end and a firefront started. The heat distils oil from the oil shales, which trickles down to the bottom of the sloping base. From here it is pumped to the surface ready for conventional refining.

Question 11

(a) Under 'normal' conditions, i.e. greater than 500 °C, carbon reacts with steam to produce carbon monoxide and hydrogen:

$$C + H_2O = CO + H_2$$

(b) Under lower temperature conditions, i.e. less than 500 °C, carbon reacts with steam to produce carbon dioxide and methane:

$$2C + 2H_2O = CO_2 + CH_4$$

The catalysts that gave most effective results at that time were alkali metals.

Question 12

It would take $10^{23}/(4 \times 10^{20})$ years, or 250 years.

Question 13

Statements (f) and (h). Statements (a) to (e) and (g) are correct. Statement (f) should read: To yield significant quantities of oil, source rocks must have a relatively high total organic carbon and be buried to depths where the temperature reaches *over 50 °C for substantial periods of time*.

Statement (h) should read: *Only 0.04% of the organic carbon preserved in sedimentary basins can be accessed for use as fuel.*

Question 14

The coal band is subdivided in Figure 83 into the following six classes:

1. Peat: decomposed fibrous plant material with high moisture content.
2. Lignite: rather soft brown coal with woody material still apparent.
3. Sub-bituminous coal: rather hard, brown coal with dull appearance developing a slight lustre.
4. Bituminous coal: hard black material with a strongly banded appearance of alternating duller and brighter bands with a shiny lustre.

5   Anthracite: very hard, black material in which banding is no longer evident and which has a bright metallic lustre.

6   Graphite: grey form of natural carbon with a silvery lustre.

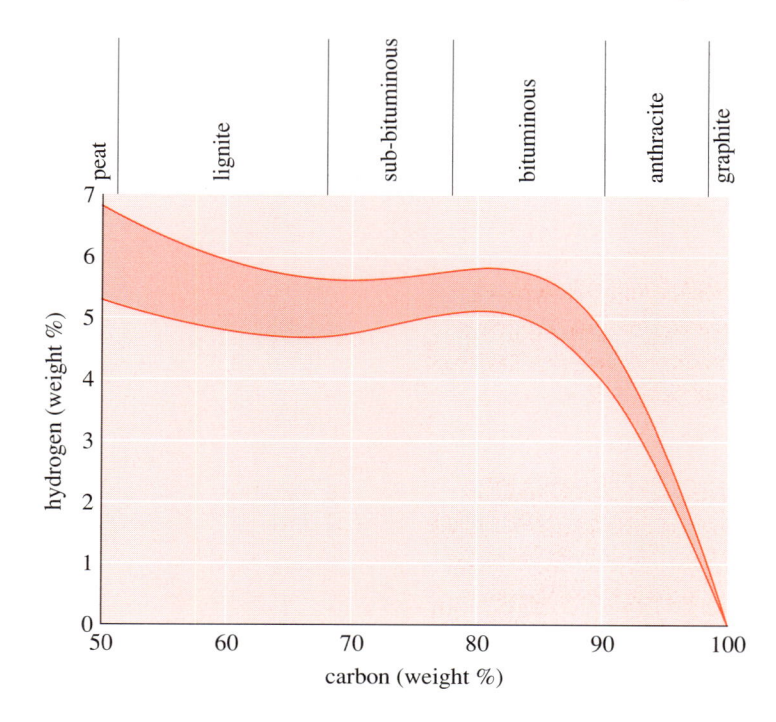

Figure 83 Answer to Question 14.

Question 15

At 125 m, the logs show low gamma and relatively high density; this cannot be coal — it is either a hard limestone or a quartz-rich sandstone. At 163 m, the logs show the low gamma–low density association typical of coal.

Question 16

The underground mining system is inflexible and needs uniform conditions to maximize its potential. This system is incapable of negotiating any serious variations in the thickness of the seam. Opencast mining is extremely flexible and can strip away all the non-coal rocks regardless of geological variations, leaving the coal free to be extracted.

Question 17

(a) A single borehole through several coal seams gives information about depth, thickness and quality of the coal seams and the nature of the inter-seam sediments.

(b) A series of boreholes through the same coal seams will allow the structure and thickness variations to be determined, reserves and stripping ratios to be calculated and hazards to be predicted.

(c) A mining engineer needs to investigate the thickness of the seam, the quality of the coal in terms of the amount of dirt and pyrite it contains and the amount of heat it can produce, and the nature of the floor in terms of load-bearing for coal-cutting machinery and roadways.

Question 18

Yes. There is no Carboniferous coal in East Anglia because it was never deposited. Figures 38 and 39 show that East Anglia was an area of low plains that were never covered by swamps at that time. Carboniferous coal was formed in the Pennines of northern England but subsequent uplift caused those rocks to be eroded.

Question 19

(a) The coal in Sparwood, British Columbia is a high-quality coal because is has been subjected to folding and faulting during formation of the Rocky Mountains. This has increased the temperature and depth of burial of the seam, and consequently increased the rank of the coal.

(b) The site at Wabamun, Alberta, is suitable for opencast working because (i) the seam is at or near the surface; (ii) at 12 m, the seam is thick, and (iii) there is only a thin overburden of unconsolidated glacial deposits, so the stripping ratio is low.

Question 20

Bands B and H are coal seams since they have both low gamma and low density values. Of the others, bands A, C, E and G are shales (high gamma, high density); band F is a marine band (very high gamma because of the high organic content); bands D and I are sandstones (higher density).

Question 21

The stripping ratio is the ratio of the thickness of barren rock that must be removed during mining to the thickness of coal extracted. The more valuable the resource, the greater the stripping ratio that can be tolerated. Anthracite commands a higher price than the bituminous coals used in power stations, so stripping ratios of about 30:1 are profitable in anthracite opencast workings, whereas bituminous coal opencasts are restricted to stripping ratios of around 20:1.

Question 22

Gradual changes (a) include seam thinning and splitting, whereas sudden changes (b) include faulting and washouts. Gradual changes result in a deterioration in the quality of coal produced because more dirt is extracted as the seam thins. Sudden changes result in the face being halted because the seam is suddenly absent.

Question 23

They all should produce good reservoirs because they are the product of processes that generate high porosity (a) and (c), or high permeability (b). Well-sorted, even-sized desert sand (a) should be highly porous. A coral reef made from a framework of animal skeletons (b) should have a large number of 'holes' between corals, and may also incidentally be a source rock — a self-sourcing reservoir. Delta sands have also been well sorted both by river transport and by marine action.

In fact, each of these geographical settings has existed in the past and produced the reservoirs of known giant oilfields in the North Sea (a and c), and America (b).

Question 24

If fractures develop in the reservoir rock *prior* to migration of oil or gas, its porosity and/or permeability would be enhanced and it would contain more petroleum. Fractures in source rocks should allow enhanced migration. However, if fractures also develop in the seal rocks, the reservoir may leak. Fluid flow through fractures can lead to the deposition of quartz or carbonate material, and seal fractures in source, reservoir and seal. This latter circumstance is good for seals and bad for source rocks and reservoirs. Predicting the presence and sealing qualities of fractures is a major part of the oil production engineer's job.

Question 25

Figure 84 shows the position of the unconformity that you should have drawn onto Figure 60b. The unconformity lies at about 2.5 s TWT at the west end of the line. It reaches a 'high' at about 2.3 s TWT at about shot point 230 and falls away eastwards to a depth of about 2.8 s TWT at shot point 500, at the east end of the line.

Figure 84 Completed version of Figure 60b.
(©1991 The Geological Society)

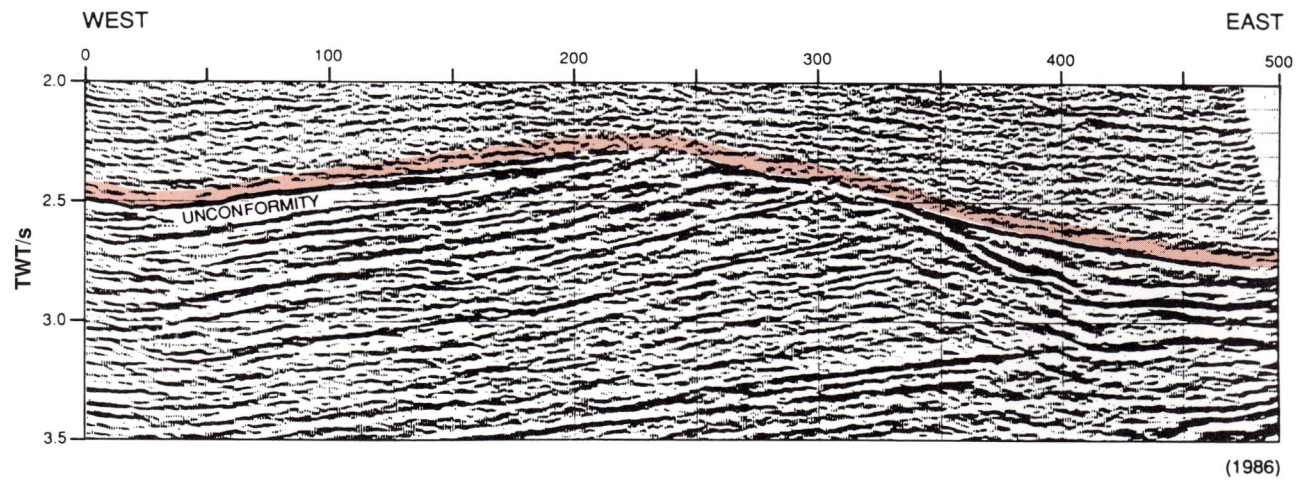

Question 26

(a) The reservoir is made from a comparatively porous reef limestone ('light brown reef') constructed by Devonian coral-like organisms called stromatoporoids.

(b) Seismic sections were of very little use in the discovery and evaluation of the Goose River oilfield because there is very little seismic velocity contrast between the reef limestones and the surrounding shales.

(c) Initially, oil was produced by *solution gas drive*, driven from the reservoir by the release of dissolved gas. The pressure drop allowed a gas cap to form, the internal pressure of which was able to drive further oil out of the reservoir. These two drives together form the primary recovery methods.

Secondary recovery was achieved by waterflood; injecting water from the surface under pressure into the reservoir and forcing oil out. Tertiary recovery is being achieved by injecting a water-detergent mixture into the reservoir, which lowers the surface tension between oil and water and breaks the oil into smaller droplets, which then is able to flow out.

(d) Primary recovery released 10–15% of the initial oil in place. Secondary recovery increased that figure to 30% (70% of the original oil was still left behind). Tertiary recovery is expected to release all but 10% of the initial oil in place, so by the end of the field's life 90% of the initial oil will have been recovered.

Question 27

Geological models based on some Bahamas reefs enabled the Goose River reservoir to be subdivided into flanking reefs surrounding a central interior lagoon. With this model, production geologists then knew that injected detergent-rich fluids would travel more readily around the fringing reef than they would pass through the impermeable interior lagoon sediments.

Question 28

Petroleum could accumulate at the following locations: 1 and 3, where the reservoir has been faulted against impermeable shale; 5, where a small anticline has formed against a fault; and a small amount at 6. At locations 2 and 4 petroleum can migrate updip, and there will be no oil at location 7 because the reservoir is open to the surface. However, an oil seepage might appear at location 7.

Question 29

The technique normally used to locate offshore traps is seismic surveying of the rocks beneath the sea bed. Drilling is the only way to determine if there is oil in the trap. From seismic sections and a programme of drilling, it is possible to determine the shapes of the trap and the reservoir. Note: gravity and magnetics cannot locate a single trap.

Question 30

The following data would be essential.

1 The volume of the reservoir and the porosity of the reservoir rock: from these, an estimate of the initial volume of reserves could be made.

2 The estimated percentage recovery. This depends on a number of things, including the nature of the reservoir rock and its permeability, and the quality of the crude oil.

3 The volume of oil already produced: subtraction of this from the estimate of the total recoverable volume would indicate how much recoverable oil was left in the field.

Question 31

(a) A water well sited at locality 1 could be used to extract water from reservoir B and inject it into reservoir A via a borehole at 2. Similarly, gas could be extracted at 4 and injected at 3 (Figure 85). Pumping water into the existing water in reservoir A would force the oil upwards; pumping gas into the gas pocket at the top of the reservoir would force the oil downwards. It might be worth considering drilling another well which intersected the gas–oil contact to tap the oil further up in the reservoir.

(b) A is a stratigraphic trap and B is a structural trap.

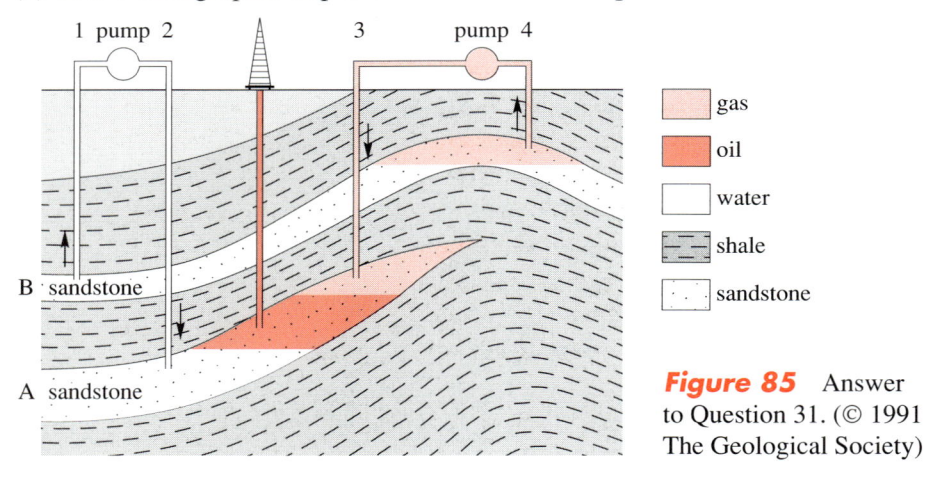

Figure 85 Answer to Question 31. (© 1991 The Geological Society)

Question 32

Ideal conditions for the generation and accumulation of petroleum prevailed in the Middle East basin. High organic productivity over a very large area led to formation of huge volumes of source rocks. Later folding and faulting produced numerous large traps into which the petroleum migrated.

Question 33

Both statements are correct, and statement (b) explains fully why statement (a) is correct. The appropriate answer from the Key is therefore E.

Question 34
Both statements are correct, but statement (b) does not provide a valid explanation for statement (a). In fact, the two statements are unrelated one to another. The appropriate answer from the Key is therefore D.

Question 35
Statement (b) is correct, but statement (a) is incorrect. Acid rain will increase as the energy requirements of the developing world get larger only if fossil fuels (particularly coal) are used to provide their energy requirements. If renewable energy sources such as solar or wind power are used, levels of acid rain will not increase. The appropriate answer from the Key is therefore C.

Question 36
Both statements are incorrect. The appropriate answer from the Key is therefore A.

Block 4

ANSWERS TO ACTIVITIES

Activity 1

(a)

(i) A kilowatt is one kilojoule per second, and there are 3600 (60×60) seconds in one hour. 1 kWh therefore represents 3600 kJ, or 3.6 MJ.

(ii) A set of bills to hand when this Activity was written shows 3057 units used in the period from 27 August to 21 February. This represents a total of 11 005 MJ (3057×3.6) for that period.

(iii) Dividing by 179 days gives 61.48 MJ each day on average.

(iv) There are 86 400 (60×60×24) seconds in a day, so the average power consumption of this household is

61 480 000/86 400 = 711.6 J s^{-1} = 711.6 W

(This is an average of £1.20 per day at 1994 prices, and also quite a shock to the householder who heats his home using gas!)

(b) Your energy consumption each day can be calculated in the same way as (a) above. The simplest way to calculate your daily power is to divide the number of units used (kWh) by the number of hours between readings. kWh/hours gives kW directly.

(c) Average electricity consumption in joules per day can vary for many reasons. If you heat your house using electricity, the outside temperature will be a significant factor. You'll use more energy on average over a cold spell, and less over a warm spell than you will on a daily basis averaged out over several weeks. Domestic situations (such as school holidays) make a difference too, when family members may be at home all day using appliances like the TV rather than being out at work. We use less domestic electricity in summer than we do in winter, since summer days are longer and summer temperatures higher. Your list will no doubt contain other relevant points that are specific to your own circumstances.

On a cold day, the length of time you may use, say, an electric fire might be the same as you would on a warmer day (the time between coming in from work and bedtime, perhaps) but you would need to use more energy to keep warm over that period on a cold day than you would on a warm day (two bars instead of one, perhaps). Accordingly, your power use is higher.

Activity 2

(a) A 2 kW rating means that the kettle has been designed to supply 2000 J every second. We want to raise the temperature of 200 g of water through 80 °C (100−20 = 80), and we know that it takes 4.18 J to raise the temperature of 1 g of water through 1 °C. We therefore need to supply 66 880 J of energy (4.18×200×80). The 2 kW kettle supplies this energy in 33.44 s (66 880/2000), i.e. just over half a minute.

(b) First, measure out 1 litre of cold tap water into a jug. 1 litre (1000 ml) will weigh 1 kg (1000 g).

We know that water boils at 100 °C, but what is the initial temperature of the water? To be accurate, you would need to measure this with a thermometer. Alternatively, if you've run the tap until water comes in from the mains, the water will be at ground temperature, which is often approximately the same as the early morning air temperature. You can get *that* information from the TV weather forecast, or use an approximate value.

Next, time how many seconds (t) it takes your kettle to just bring to the boil the litre of water starting from cold. We know it takes 4.18 J to raise the temperature of 1 g of water by 1 °C. We have 1000 g of water, and a temperature rise of 100 °C minus the initial water temperature. Call this temperature rise T °C.

A 1 kW kettle would supply 1 kJ every second, or 4.18 kJ every 4.18 s. This kettle would take 4.18 s to heat up a litre of water by 1 °C, or $T \times 4.18$ s to bring the litre of water to the boil. If the cold water was at, say, 10 °C, a 1 kW kettle would take $(100 - 10) \times 4.18 = 376$ s, or just over 6 minutes to bring a litre of water to the boil.

Now, if the kilowatt rating of your kettle were 2 kW (twice that of the 1 kW kettle), your kettle would boil the water in half the time of the 1 kW kettle. So, the time it takes a 1 kW kettle to boil divided by the time it takes your kettle to come to the boil gives you the ratio of the rating of your kettle to a 1 kW kettle:

i.e. kilowatt rating of your kettle/1 kW rating = $(T \times 4.18)/t$

therefore, the kW rating of your kettle = $(T \times 4.18)/t$ kW

I filled my kettle and performed this experiment on a day when the water temperature was 8 °C. It took 190 s to boil the litre of water. The power rating of the kettle must have been $(92 \times 4.18)/190 = 2.02$ kW. The manufacturers had in fact rated the kettle at 2 kW, but note that this experiment takes no account of energy losses — heating up the kettle, the surrounding air, and so on.

Activity 3

(a) The most likely halt could be expected from mechanical breakdowns caused by faulty or worn equipment. The shearer is powered by electricity and cooled by water sprays, so failure in the supply of either of those underground would lead to a temporary halt. Such halts are frequent occurrences underground. Despite these, the modern shearer and its operators make an efficient team; this face in Daw Mill colliery in Warwickshire yielded over 10 000 tonnes per day in 1993.

(b) Large mechanized machinery like this requires uniform geological conditions to operate efficiently. Any thinning, or sudden displacement of the seam because of past movement on faults, will halt production until the shearer can be repositioned.

Activity 4

(a) You should have been able to calculate the values in Table 6.

Table 6

d/metres	$\sqrt{d}$	$\sqrt{d} + 4$	$\dfrac{1}{\sqrt{d}+4}$	t/metres	$4t$	$\dfrac{4t}{\sqrt{d}+4} = s$/metres
50	7.07	11.07	0.090	3.0	12	1.08
150	12.25	16.25	0.062	3.0	12	0.74
250	15.81	19.81	0.050	3.0	12	0.60
350	18.71	22.71	0.044	3.0	12	0.53
450	21.21	25.21	0.040	3.0	12	0.48
550	23.45	27.45	0.036	3.0	12	0.44
650	25.49	29.49	0.034	3.0	12	0.41
750	27.39	31.39	0.032	3.0	12	0.38

and plot a graph like Figure 86.

Block 4

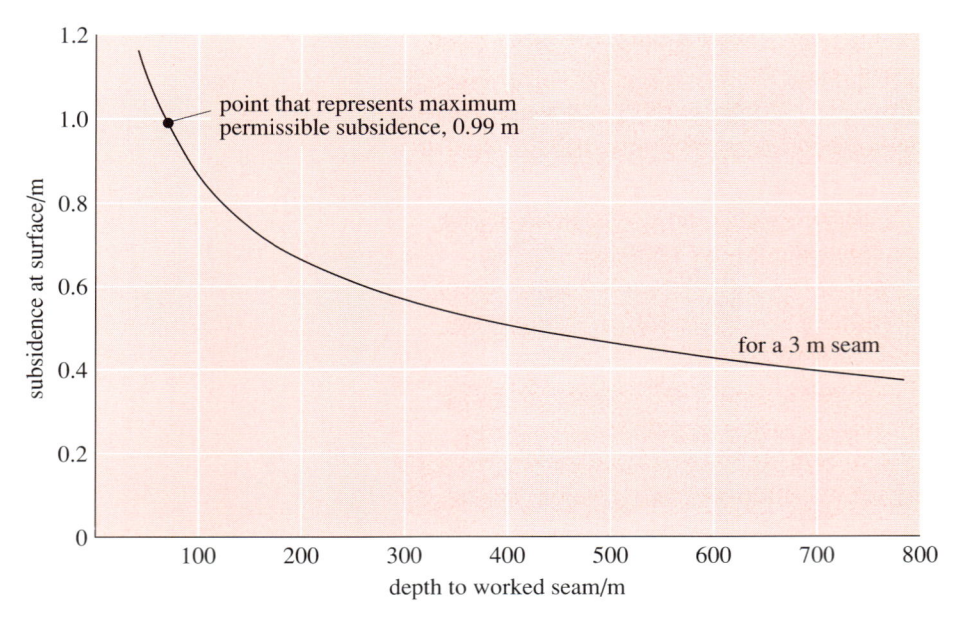

Figure 86 Graph of subsidence (vertical axis) against depth to worked seam (horizontal axis) for a seam of 3 m thickness.

(b) The graph shows that a 3 m seam can be worked at depths of about 60 m and below if it is to produce less than the 0.99 m of surface subsidence demanded at Selby.

Activity 5

(a) Each square represents 25 million tonnes of oil per year (discovered or produced) over a period of two years, in other words 50 million tonnes of oil. Since the total amount of oil discovered in UK since 1968 is represented by the area under the discovery curve between 1968 and 1992, we can estimate the area under the curve by counting squares and part-squares, and then turn this into discovery amounts using above information. UK reserves stopped increasing when the rate of production exceeded the rate of discovery. This is where the two curves cross, the year 1978.

(b) Very roughly, there are about 35 squares (sum of whole and part-squares) under the discovery curve and about 30 squares under the production curve up until 1992. By 1992, therefore, discovery amounts had exceeded production by 5 squares or 250 million tonnes of oil.

(c) Proven reserves of 4.1 billion barrels produced at 1.89 million barrels per *day* gives an R/P ratio of 2169 days or 5.9 years. Adding known reserves to hypothetical reserves (the recoverable part of the resource) gives a total of between 7.9 and 17.1 billion barrels of oil. At 1.89 million barrels per day, the higher figure of 17.1 billion barrels (best case) would take 17.1×10^9 divided by $1.89 \times 10^6 = 9048$ days, or just under 25 years to produce. This is the best-case expected reserves lifetime. Using the lower figure gives a worst-case expected reserves lifetime of 7.9×10^9 divided by 1.89×10^6, which gives just over 11 years.

Activity 6

Arild Steine and Rolf Magne Larsen note that:
- It is better to produce North Sea oil now, while it has a known value, rather than leave it in place for the future when its value is unknown.
- European Union taxes might put the cost of oil up significantly.

- Most big fields in the mature area of the North Sea have now been discovered, so North Sea oil production will decline into the next century.
- It is only profitable to exploit smaller finds in the North Sea where they are linked to the pre-existing infrastructure of a larger field.
- Drilling further north up to the ice boundary is expensive and risks damage to the environment. There have been no successful oil finds there even after 15 years of drilling.
- In the longer term energy generation will become cleaner, oil will become more taxed and gas should provide a good basis for future business.
- It is unacceptable for an oil company to face the future with a diminishing folio of oilfields and/or production possibilities.

Vidkun Hveding and Finn Lied note that:

- Oil can be bought and sold on the open market, so North Sea oil effectively 'costs' Norway the amount it can be sold for. Ownership of oilfields does not therefore mean access to cheap energy.
- An energy windfall like North Sea oil can lead to the decline of traditional industries (like fishing, in Norway's case), and create 'artificial' employment and industrial opportunities that last only as long as the oil does.
- The 50 years supply of gas should be used to generate energy while cleaner ways of burning coal are developed, but that will be costly if coal is to be used in a 'clean' way.
- If the last point is followed, coal could provide clean fuel for a further 300 years, allowing time for renewable energy to be developed. That too will be costly.
- The long term energy future lies outside fossil fuels, with nuclear power (Hveding) or solar power (Lied).
- The capital investment needed to change the basis of the world's energy supply is so large that such a change will not be accomplished for 30–50 years.
- We had better start changing now.

Your own list of main points should include most of these above, but you probably will have your own valid scenario for the two futures. Treat the second part of the answer below as one of many possible examples.

(a) Oil companies need to secure their own sellable oil and gas stocks, so that their company can stay in profit in the long term. Developing the market for North Sea gas provides some long-term security for energy supplies, but gas in a frontier area such as the Arctic will probably never be economic, or exploited. The long-term future for North Sea oil is not hopeful, for both tax and supply reasons, so the easily accessible oil should be exploited immediately. In the longer term, the future of oil lies outside the North Sea area.

(b) Politicians tend to look towards activities that can be achieved in the short term. Raising oil taxes could bring in revenue, reduce CO_2 emissions and extend the lifetime of indigenous oil supplies. Moving towards gas as an electricity generator makes political sense; it is cleaner and longer lasting than oil and much cleaner than coal. Investing in renewable energy supplies is not demanded yet by their electorate, so whereas it seems like a good idea to retired politicians (who often can take a long-term, strategic view of events), it is unlikely to be acted on by serving politicians looking towards the next election.

Activity 7

(a) The countries which have the highest fallout of sulphur (over $6\,g\,m^{-2}$) include England, Germany, Poland, Czech Republic, Slovakia, Croatia, Ukraine, Russia, Belgium, Austria, Switzerland and north Italy. The countries which have the lowest fallout (under $1\,g\,m^{-2}$) include Portugal, much of Spain, Republic of Ireland, most of Norway, Sweden and Finland, and most of Turkey.

(b) Most of the countries with the highest levels of sulphur precipitation either have extensive Carboniferous coal reserves (Figure 39) or extensive Tertiary lignite reserves (Figure 41). Given that coal has a high place value, at least in historical terms, it is reasonable to assume that these countries still generate most of their electricity in solid-fuel power stations. Conversely, countries without any indigenous coal, like Iceland, Norway and Sweden have very much lower sulphur precipitation rates.

(c) Ireland has such a low fallout, despite generating much of its electricity from solid fuel, because it lies on the western fringe of Europe, upwind from most coal-fired power stations given north-west Europe's prevailing westerly winds. Its own sulphur fallout will tend to drift downwind towards England. Denmark and southern Sweden have high fallout amounts despite having no coal because they lie downwind of the big coal users of Britain, Germany and Poland.

(d) There is no unique answer to this part of the Activity. Since sulphur precipitation seems to be linked to coal and lignite, reducing the sulphur emissions from the flues of coal-fired power stations in Britain, Germany and Eastern Europe should help. Sulphur emissions might also be associated with car use, so increased provision and use of electric trains in most countries could help. Scrubbing all flue gases should reduce SO_2 drastically.

Acknowledgements

The author would like to thank the Block Assessor, Eric Skipsey, for his helpful comments and suggestions on the earlier drafts. The following student readers are thanked for their comments on an early draft: Julia Adamson, Tom Denne and Iris Rowbotham.

Grateful acknowledgement is made to the following sources for permission to reproduce material in this block.

Figures

Cover: Satellite composite view of Earth, copyright © 1990 Tom Van Sant/The GeoSphere® Project, Santa Monica, California, with assistance from NOAA, NASA, EYES ON EARTH, technical direction Lloyd Van Warren, source data derived from NOAA/TIROS-N Series Satellites. All rights reserved; *Figure 8:* British Gas; *Figure 12:* 'A diorama at the Geological Museum showing a Carboniferous coal forest', The Natural History Museum, London; *Figure 13:* Skinner, B. J., *Earth Resources*, ®1976, figure 13. Adapted by permission of Prentice-Hall, Inc. Englewood Cliffs, NJ; *Figures 16 and 69:* Tarling, D. H. (1981), *Economic Geology and Tectonics*, Blackwell Science Ltd; *Figure 17:* Hunt, J. M., *Geochemistry of Petroleum*, American Association of Petroleum Geologists; *Figure 18:* Day, G. A. *et al.* (1981), in Illing, L. V. and Hobson, G. D. (eds), *Regional Seismic Structure Maps of the North Sea*, John Wiley and Sons Ltd. Reprinted by permission of John Wiley and Sons Ltd; *Figure 19:* Tissot, B. P. and Welte, D. H. (2nd edition 1978) 'From kerogen to petroleum' in *Petroleum Formation and Occurrence*, © by Springer-Verlag Berlin Heidelburg 1978 and 1984; *Figure 20:* Ziegler, P. A. (1993) 'Plate moving mechanisms: their relative importance', in Lebas, M. J. (ed.), *Journal of the Geological Society*, London, 150, 1993, pp. 927–940, © 1993 The Geological Society; *Figures 21 and 83:* Mason, B. (3rd edition 1966), *Principles of Geochemistry*, John Wiley and Sons Ltd. Reprinted by permission of John Wiley and Sons Ltd; *Figure 23:* Selley, R. C. (1985), *Elements of Petroleum Geology*, R. C. Selley and Company Ltd; *Figure 24:* Exploration Logging, Inc.; *Figure 25:* Reeves, D. R. (1978) 'Some improvements and developments in coal wireline logging techniques', in Argall, G. A. (ed.), *Coal Exploration 2*, Proceedings of the 2nd International Coal Exploration Symposium, Denver, October 1978, pp. 112–128, Miller-Freeman, San Francisco; *Figure 26:* Royal Commission on the Historical Monuments of England, National Monuments Record, Air Photographs, © Crown Copyright. Reproduced with the permission of the Controller of Her Majesty's Stationery Office; *Figure 27:* Cossons, N. (2nd edition 1987) 'Coal', in *The BP Book of Industrial Archaeology*, David and Charles Publishing plc; *Figure 28:* Popperfoto; *Figures 30 and 31:* Guion, P. D. and Fulton, I. M. (1993) 'The importance of sedimentology in deep-mined coal extraction', in *Geoscientist*, 3(2), March/April 1993, © 1993 The Geological Society; *page 59:* Science Museum/Science & Society Picture Library; *Figure 36:* The Selby Coalfield, British Coal; *Figures 38 and 41:* Ziegler, P. A. (1987), *Evolution of the Arctic: North Atlantic and the Western Tethys*, © Shell Internationale Petroleum Maatschappij N. V.; *Figure 39:* Ziegler, P. A. (1987), *Evolution of the Arctic: North Atlantic and the Western Tethys*, Copyright © 1988 The American Association of Petroleum Geologists; *Figure 40:* Francis, E. H. (1979) 'British Coalfields' in *Science Progress*, 66, Science Technology Letters; *Figure 42:* based on a map by Prof. A. J. Smith; *Figure 43:* 'Coal', in *BP Statistical Review of World Energy*, June 1993, © The British Petroleum Company Plc 1993; *Figure 57:* Stäuble, A. J. and Milius, G.

(1970) 'The geology of the Groningen Gas Field, Netherlands', in Halbouty, M. T. (ed), *Geology of Giant Petroleum Fields*, Copyright © 1970 by The American Association of Petroleum Geologists; *Figure 58:* The Natural History Museum, London; *Figures 60, 63, 84 and 85:* Struijk, A. P. and Green, R. T. (1991) 'The Brent Field, 211/29, UK North Sea', in Abbotts, I. L. (ed), *United Kingdom Oil and Gas Fields*, Commemorative Volume, Geological Society Memoir No.14, pp. 63–72, © 1991 The Geological Society; *Figure 61:* Glenbow Archives, Calgary, Alberta, Photo no. NA 1585–3; *Figures 62 and 64b:* Woodcock, N. H. (1994), *Geology and Environment in Britain and Ireland*, UCL Press Limited; © Nigel H. Woodcock 1994; *Figure 64a: Annual Review of Energy*, US Department of Energy; *Figure 66:* Parsley, A. J. (1990), 'North Sea hydrocarbon plays' in Glennie, K. W. (ed.), *Introduction to the Petroleum Geology of the North Sea*, Blackwell Science Ltd; *Figure 67:* Coinford (1990), *Introduction to Petroleum Geology of North Sea*, Blackwell Science Ltd; *Figure 68:* 'Oil', in *BP Statistical Review of World Energy*, June 1993, © The British Petroleum Company Plc 1993; *Figure 69: Figure 70:* 'Gas', in *BP Statistical Review of World Energy*, June 1993, © The British Petroleum Company plc 1993; *Figures 77 and 78:* Pearce, F. (1987) 'Acid rain', in *New Scientist*, 116(1585), 5 November 1987, © IPC Magazines, 1987; *Figure 79:* Brown, A. (1993), *The UK Environment*, © Crown Copyright. Reproduced with the permission of the Controller of Her Majesty's Stationery Office; *Figure 81:* Barnola, J. M., Raynaud, D., Lorius, C. and Korotkevich, Y. S. (1987) 'Vostok ice core provides 160,000 year record of atmospheric CO_2, Reprinted with permission from *Nature*, 329, p.410, 1 October 1987, Copyright © 1987 Macmillan Magazines Limited.

Block 4

Physical Resources and Environment

Block 1   *Physical Resources – an Introduction*
Block 2   *Building Materials*
Block 3   *Water Resources*
Block 4   *Energy 1 – Fossil Fuels*
Block 4   *Energy 2 – Nuclear and Alternatives*
Block 5   *Metals 1 – Ore Deposits*
Block 5   *Metals 2 – Resource Exploitation*
Block 6   *Topics and Data*